Applications of Fibonacci Numbers

Applications of Fibonacci Numbers

Proceedings of 'The Second International Conference on
Fibonacci Numbers and Their Applications'
San Jose State University, California, U.S.A.
August 1986

edited by

A. N. Philippou
Department of Mathematics,
University of Patras, Greece

A. F. Horadam
Department of Mathematics, Statistics and Computer Science,
University of New England, Armidale, Australia

and

G. E. Bergum
Computer Science Department,
South Dakota State University, Brookings, U.S.A.

KLUWER ACADEMIC PUBLISHERS

DORDRECHT / BOSTON / LONDON

0304-7933

MATH-STAT.

Library of Congress Cataloging in Publication Data

CIP

Applications of Fibonacci numbers / edited by A. N. Philippou, A. F. Horadam and
    G. E. Bergum.
        p.   cm.
    "Papers presented at the Second International Conference on Fibonacci Numbers and
Their Applications"—Foreword.
    "Co-sponsored by the Fibonacci Association and San Jose State University"—P. ix.
    Includes bibliographies and index.
    ISBN 90–277–2673–6
    1.   Fibonacci numbers—Congresses.   I.   Philippou, Andreas N.
II.   Horadam, A. F.   III.   Bergum, Gerald E.   IV.   International Conference on
Fibonacci Numbers and Their Applications (2nd: 1986: San Jose State University).
V.   Fibonacci Association.   VI.   San Jose State University.
QA241.A66   1987
512'.72—dc 19

87–37648
CIP

Published by Kluwer Academic Publishers,
P.O. Box 17, 3300 AA Dordrecht, The Netherlands.

Kluwer Academic Publishers incorporates
the publishing programmes of
D. Reidel, Martinus Nijhoff, Dr W. Junk and MTP Press.

Sold and distributed in the U.S.A. and Canada
by Kluwer Academic Publishers,
101 Philip Drive, Norwell, MA 02061, U.S.A.

In all other countries, sold and distributed
by Kluwer Academic Publishers Group,
P.O. Box 322, 3300 AH Dordrecht, The Netherlands.

TABLE OF CONTENTS

THE SECOND INTERNATIONAL...   vii
CONTRIBUTORS   ix
FOREWORD   xiii
THE ORGANIZING COMMITTEES   xv
LIST OF CONTRIBUTORS TO THE CONFERENCE   xvii
INTRODUCTION   xix

FERMAT-LIKE BINOMIAL EQUATIONS
    Heiko Harborth 1

RECURRENCES RELATED TO THE BESSEL FUNCTION
    F. T. Howard 7

SYMMETRIC RECURSIVE SEQUENCES MOD M
    Kenji Nagasaka & Shiro Ando 17

PRIMITIVE DIVISORS OF LUCAS NUMBERS
    Peter Kiss 29

A CONGRUENCE RELATION FOR A LINEAR RECURSIVE SEQUENCE OF
ARBITRARY ORDER
    H. T. Freitag & G. M. Phillips 39

FIBONACCI NUMBERS AND GROUPS
    Colin M. Campbell, Edmund F. Robertson
    & Richard M. Thomas 45

A TRIANGULAR ARRAY WITH HEXAGON PROPERTY, DUAL TO PASCAL'S
TRIANGLE
    Shiro Ando 61

FUNCTIONS OF THE KRONECKER SQUARE OF THE MATRIX Q
    Odoardo Brugia & Piero Filipponi 69

FIBONACCI NUMBERS OF THE FORMS $PX^2 \pm 1$, $PX^3 \pm 1$, WHERE P
IS PRIME
    Neville Robbins 77

ON THE K-TH ORDER LINEAR RECURRENCE AND SOME PROBABILITY
APPLICATIONS
    George N. Philippou 89

ON THE REPRESENTATION OF INTEGRAL SEQUENCES $\{F_n/d\}$ and
$\{L_n/D\}$ AS SUMS OF FIBONACCI NUMBERS AND AS SUMS OF LUCAS
NUMBERS
  Herta T. Freitag & Piero Filipponi   97

PRIMES HAVING AN INCOMPLETE SYSTEM OF RESIDUES FOR A
CLASS OF SECOND-ORDER RECURRENCES
  Lawrence Somer   113

COVERING THE INTEGERS WITH LINEAR RECURRENCES
  John R. Burke & Gerald E. Bergum   143

RECURSIVE THEOREMS FOR SUCCESS RUNS AND RELIABILITY OF
CONSECUTIVE-K-OUT-OF-N: F SYSTEMS
  Andreas N. Philippou   149

ASVELD'S POLYNOMIALS $P_j(N)$
  A. F. Horadam & A. G. Shannon   163

MORE ON THE PROBLEM OF DIOPHANTUS
  Joseph Arkin & Gerald Bergum   177

ON A PROBLEM OF DIOPHANTUS
  Calvin Long & Gerald Bergum   183

THE GENERALIZED FIBONACCI NUMBERS $\{C_n\}$, $C_n = C_{n-1} + C_{n-2} + K$
  Marjorie Bicknell-Johnson & Gerald E. Bergum   193

FIRST FAILURES
  Dmitri Thoro   207

SUBJECT INDEX   211

THE SECOND INTERNATIONAL CONFERENCE
ON
FIBONACCI NUMBERS AND THEIR APPLICATIONS

A MEMORY LADEN EXPERIENCE

There I was--alone on a strange campus, at the University of California at Berkeley, where the startling number of 3,970 had gathered for ICM-86, The International Congress of Mathematicians. Did someone just call my name? He had done it again!--Professor A. N. Philippou, Chairman of our First International Conference on Fibonacci Numbers and Their Applications two years ago at The University of Patras, Greece, the man who at the time had "recognized" me without ever having seen me, now managed to "run into me" amidst this "almost non-denumerable" crowd.

To encounter-just before our Conference-Professor Philippou, the originator of the idea to set the stage for the meeting of "Fibonacci friends" on an international scale, was a very special omen to me. It was an appropriate and beautiful overture to our Second International Conference on Fibonacci Numbers and Their Applications, which was to begin two days later, and convened from August 13-16 at San Jose State University. This site was befittingly chosen as it is the home of *The Fibonacci Quarterly*.

Professor Calvin Long, Chairman of the Board of The Fibonacci Association, and Professor Hugh Edgar, a member of the University's Mathematics Department, participated in the Conference. This gave us the opportunity to express our appreciation of the fact that our Conference was co-sponsored by The Fibonacci Association and San Jose State University.

Professor Gerald E. Bergum, Editor of *The Fibonacci Quarterly* and Chairman of the Local Committee, and Professor A. N. Philippou, who chaired the International Committee, immediately earned our admiration and praise. So did the Co-Chairmen-Professors A. F. Horadam and Hugh Edgar, and, indeed, Professor Calvin Long and all the other helpers "on the stage" and "in the wings."

The organization of our Conference was exemplary. And the atmosphere was charged with that most appealing blend of the seriousness and profundity of scholarliness and the enthusiasm and warmth of personal relationships. This seems to be the trademark of "Fibonaccians"--mathematicians who are dedicated to a common cause: a deep and abiding fascination with "Fibonacci-type" mathematics.

Approximately twenty-five papers were presented by a group which came from some ten countries. There were several joint authorships. Some had resulted from a cooperation between authors separated by oceans-a situation which, predictably, poses many obstacles: one just has to "hover by the mailbox until the anxiously awaited response can possibly arrive." Many of the papers exhibited the

vii

phenomenon that one mathematical idea begot another, and yet another, maybe a generalization, and yet a further one, etc., the very development mathematicians cherish so much. Our understanding of the goldmine that number sequences and the intricacies of their interrelationships constitute was enriched, and our appreciation of the value of such investigations was deepened. While the variety of topics was striking, dedication to the beauty of mathematical patterns and joy over the wealth of mathematical relationships provided the common bond. This book is a culmination of the papers presented at the Conference.

A small nucleus, just seven participants, were "second-timers," people who had previously experienced the unique pleasure of this kind of gathering on an international scale. Their friendships were welded together more meaningfully yet, and many newcomers were initiated. Many of us had accents but, in a very significant way, we all spoke the same language.

Professor Hoggatt's widow, Herta Hoggatt, most graciously invited our entire mathematical communtiy to convene at her charming home-outdoors, amidst the beauty of flowers and trees. In a deeply touching way did the late Professor Verner E. Hoggatt, Jr., thus participate in our thoughts.

I believe that all of our Fibonacci friends--here and across the oceans--greatly valued the fact that the dream, first voiced in Greece, about continuation of our international gatherings had been realized. Now, we confidently rejoice over the prospect: "Until we meet again..., in two years, in Italy..., maybe in Pisa!"

Herta T. Freitag

CONTRIBUTORS

PROFESSOR SHIRO ANDO (pp. 17-28; 61-67)
College of Engineering
Hosei University
3-7-2, Kajino-Cho
Koganei-Shi, Tokyo 184
JAPAN

MR. JOSEPH ARKIN (pp. 177-181)
197 Old Nyack Turnpike
Spring Valley, NY 10977 U.S.A.

PROFESSOR GERALD E. BERGUM (pp. 143-147; 177-181; 183-191; 193-205)
Computer Science Department
South Dakota State University
Box 2201
Brookings, SD 57007-0194 U.S.A.

PROFESSOR ODOARDO BRUGIA (pp. 69-76)
Fondazione Ugo Bordoni
Viale Trastevere, 108
00153-Roma, ITALY

PROFESSOR JOHN BURKE (pp. 143-147)
Department of Math. and Comp. Sci.
Gonzaga University
Spokane, WA 99258 U.S.A.

PROFESSOR COLIN CAMPBELL (pp. 45-60)
The Mathematical Institute
University of St. Andrews
The North Haugh
St. Andrews KY 16 9SS
Fife, SCOTLAND

PROFESSOR PIERO FILIPPONI (pp. 69-76; 97-112)
Fondazione Ugo Bordoni
Viale Trastevere, 108
00153-Roma ITALY

PROFESSOR HERTA FREITAG   (pp. 39-44; 97-112)
B-40 Friendship Manor
320 Hershberger Road, N.W.
Roanoke, VA   24012   U.S.A.

DR. HEIKO HARBORTH   (pp. 1-5)
Bienroder Weg 47
D-3300 Braunschweig
WEST GERMANY

PROFESSOR A. F. HORADAM   (pp. 163-176)
Dept. of Math., Stat., & Comp. Sci.
University of New England
Armidale N.S.W. 2351
AUSTRALIA

PROFESSOR FRED T. HOWARD   (pp. 7-16)
Department of Mathematics
Wake-Forest University
Winston-Salem, NC   27109   U.S.A.

DR. MARJORIE BICKNELL-JOHNSON   (pp. 193-205)
665 Fairlane Avenue
Santa Clara, CA   95051   U.S.A.

PROFESSOR PETER KISS   (pp. 29-38)
3300 Eger
Csiky S.U. 7 mfsz. 8
HUNGARY

PROFESSOR CALVIN LONG   (pp. 183-191)
Department of Mathematics
Washington State University
Pullman, WA   99163   U.S.A.

PROFESSOR KENJI NAGASAKA   (pp. 17-28)
University of the Air
2-11, Wakaba, Chiba-Shi
260, Chiba, JAPAN

PROFESSOR ANDREAS N. PHILIPPOU   (pp. 149-161)
Department of Mathematics
University of Patras
Patras, GREECE

PROFESSOR GEORGE N. PHILIPPOU   (pp. 89-96)
Higher Technical Institute
P.O. Box 2423
Nicosia CYPRUS

PROFESSOR GEORGE M. PHILLIPS (pp. 39-44)
The Mathematical Institute
The North Haugh
St. Andrews, SCOTLAND KY 16 9SS

PROFESSOR NEVILLE ROBBINS (pp. 77-88)
Department of Mathematics
1600 Holloway Avenue
San Francisco State University
San Francisco, CA 94132 U.S.A.

PROFESSOR EDMUND F. ROBERTSON (pp. 45-60)
Mathematical Institute
University of St. Andrews
St. Andrews SCOTLAND KY 16 9SS

PROFESSOR A. G. SHANNON (pp. 163-176)
The New South Wales Institute of Technology
School of Mathematical Sciences
P.O. Box 123
Broadway N.S.W. 2007
AUSTRALIA

PROFESSOR LAWRENCE SOMER (pp. 113-141)
1400 20th St., NW #619
Washington, DC 20036 U.S.A.

PROFESSOR RICHARD M. THOMAS (pp. 45-60)
Department of Mathematics
St. Mary's College
Twickenham ENGLAND TW1 4SX

PROFESSOR DMITRI THORO (pp. 207-210)
Department of Math. and Comp. Sci.
San Jose State University
San Jose, CA 95182-0103 U.S.A.

FOREWORD

This book contains nineteen papers from among the twenty-five papers presented at the Second International Conference on Fibonacci Numbers and Their Applications. These papers have been selected after a careful review by well known referee's in the field, and they range from elementary number theory to probability and statistics. The Fibonacci numbers are their unifying bond.

It is anticipated that this book will be useful to research workers and graduate students interested in the Fibonacci numbers and their applications.

October 1987   The Editors

Gerald E. Bergum
South Dakota State University
Brookings, South Dakota, U.S.A.

Andreas N. Philippou
University of Patras
Patras, Greece

Alwyn F. Horadam
University of New England
Armidale, N.S.W., Australia

THE ORGANIZING COMMITTEES

LOCAL COMMITTEE

Bergum, G., *Chairman*

Edgar, H., *Co-chairman*

Thoro, D.

Johnson, M.

Lange, L.

INTERNATIONAL COMMITTEE

Philippou, A. (Greece), *Chairman*

Horadam, A. (Australia), *Co-chairman*

Bergum, G. (U.S.A.)

Kiss, P. (Hungary)

Long, C. (U.S.A.)

Ando, S. (Japan)

LIST OF CONTRIBUTORS TO THE CONFERENCE*

*ANDO, S., Hosei University, Tokyo (coauthor K. Nagasaka). "Symmetric Recursive
   Sequences Mod m."
*ANDO, S., Hosei University, Tokyo. "A Triangular Array with Hexagon Property
   Which is Dual to Pascal's Triangle."
*ARKIN, J., Spring Valley, New York (coauthor G. E. Bergum) "More on the
   Problem of Diophantus."
*BERGUM, G. E., South Dakota State University, Brookings, South Dakota,
   (coauthor M. Johnson). The Generalized Fibonacci Numbers
   {C_n}, $C_n = C_{n-1} + C_{n-2} + K$."
*BERGUM, G. E., South Dakota State University, Brookings, South Dakota,
   (coauthor J. Arkin). "More on the Problem of Diophantus."
*BERGUM, G. E., South Dakota State University, Brookings, South Dakota,
   (coauthor J. Burke). "Covering the Integers With Linear Recurrences."
*BERGUM, G. E., South Dakota State University, Brookings, South Dakota,
   (coauthor C. Long). "Linear Recurrences and the Problem of Diophantus."
*BRUGIA, O., Fondazione Ugo Bordoni, Roma (coauthor P. Filipponi). "Functions of
   the Kronecker Square of the Matrix Q."
*BURKE, J., Gonzaga University, Spokane, Washington (coauthor G. E. Bergum).
   "Covering the Integers With Linear Recurrences."
*CAMPBELL, C. M., University of St. Andrews, St. Andrews (coauthors E. F.
   Robertson and R. M. Thomas). "Fibonacci Numbers and Groups."
 CLARKE, J. H., New South Wales Institute of Technology, Sidney (coauthors
   A. G. Shannon and L. J. Hills). "Contingency Relations for Infectious Diseases."
*FILIPPONI, P., Fondazione Ugo Bordoni, Roma (coauthor O. Brugia). "Functions of
   the Kronecker Square of the Matrix Q."
*FILIPPONI, P., Fondazione Ugo Bordoni, Roma (coauthor H. Freitag) "On the
   Representation of Integral Sequences {F_n/d} and {L_n/d} as Sums of Fibonacci
   Numbers and as Sums of Lucas Numbers."
*FREITAG, H., Roanoke, Virginia. "On the Representation of Integral Sequences
   {F_n/d} and {L_n/d} as Sums of Fibonacci Numbers and as Sums of Lucas Numbers."
*FREITAG, H., Roanoke, Virginia. (coauthor G. N. Phillips) "A Congruence Relation
   for a Linear Recursive Sequence of Arbitrary Order."
*HARBORTH, H., Bienroder Weg 47, West Germany. "Fermat-Like Binomial
   Equations."

----

* The asterisk indicates that the paper is included in this book.

HILLS, L. J., New South Wales Institute of Technology, Sidney (coauthors A. G. Shannon and J. H. Clarke). "Contingency Relations for Infectious Diseases."

*HORADAM, A. F., University of New England, Armidale (coauthor A. G. Shannon). "Asveld's Polynomials $P_j(n)$."

HORADAM, A. F., University of New England, Armidale (coauthor A. P. Treweek). "Simson's Formula and an Equation of Degree 24."

*HOWARD, F., Wake-Forest University, Winston-Salem, North Carolina. "Recurrences Related to the Bessel Function."

*JOHNSON, M. Santa Clara, California, (coauthor G. E. Bergum). "The Generalized Fibonacci Numbers $\{C_n\}$, $C_n = C_{n-1} + C_{n-2} + K$."

*KISS, P., 3300 Eger, Csiky. "Primitive Divisors of Lucas Numbers."

*LONG, C., Washington State University, Pullman, Washington, (coauthor G. E. Bergum). "Linear Recurrences and the Problem of Diophantus."

MOLLIN, R., University of Calvary, Alberta. "Generalized Primitive Roots."

*NAGASAKA, K., University of the Air, Chiba (coauthor S. Ando). "Symmetric Recursive Sequences Mod m."

NIEDERREITER, H., Austrian Academy of Sciences, Vienna (coauthor P. J. Shuie). "Weak Uniform Distributions of Linear Recurring Sequences in Finite Fields."

*PHILIPPOU, A. N., University of Patras, Patras. "Recursive Theorems for Success Runs."

PHILIPPOU, A. N., University of Patras, Patras. "A Closed Formula for Strict Consecutive-k-out-of-n: F Systems."

*PHILIPPOU, G. N., Higher Technical Institute. "On the k-th Order Linear Recurrence and Some Probability Applications."

*PHILLIPS, G. N., The Mathematical Institute, St. Andrews (coauthor H. Freitag). "A Congruence Relation for a Linear Recursive Sequence of Arbitrary Order."

*ROBBINS, N., San Francisco State University, San Francisco, California. "Fibonacci Numbers of the Forms $px^2 \pm 1$, Where p is Prime."

*ROBERTSON, E. F., University of St. Andrews, St. Andrews (coauthors C. M. Campbell and R. M. Thomas). "Fibonacci Numbers and Groups."

SATO, D., University of Regina, Regina. "Star of David Theorem (II)-A Simple Proof of Ando's Theorem."

SHANNON, A. G., The New South Wales Institute of Technology, Broadway (coauthors J. H. Clarke and L. J. Hills). "Contingency Relations for Infectious Diseases."

*SHANNON, A. G., The New South Wales Institute of Technology, Broadway (coauthor A. F. Horadam). "Asveld's Polynomials $P_j(n)$."

SHUIE, P. J., University of Nevada, Las Vegas (coauthor H. Niederreiter) "Weak Uniform Distributions of Linear Recurring Sequences in Finite Fields."

*SOMER, L., Washington, D.C. "Primes Having Incomplete Residue Systems for a Class of Second-Order Recurrences."

SPIRO, C., University of Buffalo, Buffalo. "A Set Generated by a Linear Recurrence and a Dot Product Rule."

*THOMAS, R. M., St. Mary's College, Twickenham (coauthors C. M. Campbell and E. F. Robertson). "Fibonacci Numbers and Groups."

*THORO, D., San Jose State University, San Jose, California. "First Failures."

TREWEEK, A. P., Sidney, Australia (coauthor A. F. Horadam). "Simson's Formula and an Equation of Degree 24."

INTRODUCTION

The numbers

   1, 1, 2, 3, 5, 8, 13, 21, 34, 55, 89, . . . ,

known as Fibonacci numbers have been named by the nineteenth-century French mathematician Edouard Lucas after Leonard Fibonacci of Pisa, one of the best mathematicians of the Middle Ages, who referred to them in his book *Liber Abaci* (1202) in connection with his rabbit problem.

   The astronomer Johann Kepler rediscovered the Fibonacci numbers, independently, and since then several renowned mathematicians have dealt with them. We only mention a few: J. Binet, B. Lamé, and E. Catalan. Edouard Lucas studied Fibonacci numbers extensively, and the simple generalization

   2, 1, 3, 4, 7, 11, 18, 29, 47, 76, 123, . . . ,

bears his name.

   During the twentieth century, interest in Fibonacci numbers and their applications rose rapidly. In 1961 the Soviet mathematician N. Vorobyov published *Fibonacci Numbers*, and Verner E. Hoggatt, Jr., followed in 1969 with his *Fibonacci and Lucas Numbers*. Meanwhile, in 1963, Hoggatt and his associates founded The Fibonacci Association and began publishing *The Fibonacci Quarterly*. They also organized a Fibonacci Conference in California, U.S.A., each year for almost sixteen years until 1979. In 1984, the First International Conference on Fibonacci Numbers and Their Applications was held in Patras, Greece, and the proceedings from this conference have been published. It was anticipated at that time that this conference would set the beginning of international conferences on the subject to be held every two or three years in different countries. The Second International Conference on Fibonacci Numbers and Their Applications was held in San Jose, California, U.S.A., August 13-16, 1986, which this book is the result of. The Third International Conference on Fibonacci Numbers and Their Applications is planned to take place in Pisa, Italy, in late July of 1988.

It is impossible to overemphasize the importance and relevance of the Fibonacci numbers to the mathematical sciences and other areas of study. The Fibonacci numbers appear in almost every branch of mathematics, like number theory, differential equations, probability, statistics, numerical analysis, and linear algebra. They also occur in biology, chemistry, and electircal engineering.

It is believed that the contents of this book will prove useful to everyone interested in this important branch of mathematics and that they may lead to additional results on Fibonacci numbers both in mathematics and in their applications in science and engineering.

The Editors

Heiko Harborth

FERMAT-LIKE BINOMIAL EQUATIONS

For more than three centuries it has been a conjecture that the Diophantine equation

$$x^n + y^n = z^n \tag{1}$$

has no solutions in natural numbers x, y, z for all n > 2. At present this conjecture, which is also called "Fermat's Last Theorem", is known to be true for all $n \leq 125\ 000$ [1]. Moreover, the recent work of G. Faltings (see [1]) implies that, for each $n \geq 3$, (1) has at most a finite number of solutions (x, y, z), with (x, y, z) = 1 and $xyz \neq 0$.

In this paper, instead of n'th powers being considered, binomial coefficients within a row n, within a column n, or within a Fibonacci row n of Pascal's triangle are considered, where a Fibonacci row n denotes all numbers $\binom{n-i}{i}$, $0 \leq i \leq \frac{n}{2}$, since their sum equals the Fibonacci number F_{n+1} with $F_0 = 0$, $F_1 = 1$, and $F_{n+2} = F_{n+1} + F_n$. In other words, solutions in natural numbers n, x, y, z of the following three equations are sought:

$$\binom{n}{x} + \binom{n}{y} = \binom{n}{z}, \tag{2}$$

$$\binom{x}{n} + \binom{y}{n} = \binom{z}{n}, \tag{3}$$

$$\binom{n-x}{x} + \binom{n-y}{y} = \binom{n-z}{z}. \tag{4}$$

Of course, for fixed n in (2) and (4) there are only finitely many triples (x, y, z) possible. In the following, for each of the equations (2), (3), and (4), solutions (x, y, z) are given for infinitely many values of n. If $\binom{n}{k}$ solves one of the equations, then $\binom{n}{n-k}$ is used in the case when n - k < k.

1

A. N. Philippou et al. (eds.), Applications of Fibonacci Numbers, 1–5.
© 1988 by Kluwer Academic Publishers.

Theorem 1: Equation (2) has at least the following solutions:

$$(x, y, z) = (x, x, x + 1) \text{ for } n = 3x + 2 \quad (x = 0, 1, 2, \ldots), \tag{1.1}$$

$$(x, y, z) = (x, x, x + 2) \text{ for } n = n_i, \; x = x_i \quad (i = 1, 2, \ldots) \text{ with} \tag{1.2}$$
$$n_1 = 5, \; x_1 = 1, \; n_{i+1} = 5n_i + 2x_i + 7, \text{ and } x_{i+1} = 2n_i + x_i + 2,$$

$$(x, y, z) = (x, x + 1, x + 2) \text{ for } n = F_{2i+2}F_{2i+3} - 1, \text{ and} \tag{1.3}$$
$$x = F_{2i}F_{2i+3} - 1 \quad (i = 1, 2, \ldots).$$

Proof: If $x = y = z - 1$ is substituted in (2), then $2\binom{n}{x} = \binom{n}{x+1}$, and this is equivalent to $n = 3x + 2$, so that (1.1) follows. In the case (1.2) the substitution of $x = y = z - 2$ in (2) implies $2\binom{n}{x} = \binom{n}{x+2}$, which is equivalent to $2(x+1)(x+2) = (n-x)(n-x-1)$, and this yields

$$x = \tfrac{1}{2}\left(-2n - 5 + \sqrt{2(2n+2)^2+1}\right). \tag{1.4}$$

Now all solutions of $s^2 - 2t^2 = 1$ in natural numbers are known to be (see [2], p. 94)

$$(s_1, t_1) = (3, 2), \quad (s_{i+1}, t_{i+1}) = (3s_i+4t_i, 2s_i+3t_i) \tag{1.5}$$

with $i = 1, 2, \ldots$. For $n = n_i$ and $x = x_i$ equation (1.4) is fulfilled by n_i and x_i from $s_i = 2x_i + 2n_i + 5$, and $t_i = 2n_i + 2$. The recursions of (1.5) yield

$$s_{i+1} = 2x_{i+1} + 2n_{i+1} + 5 = 3(2x_i + 2n_i + 5) + 4(2n_i + 2),$$
$$t_{i+1} = 2n_{i+1} + 2 = 2(2x_i + 2n_i + 5) + 3(2n_i + 2).$$

Then by elimination of n_{i+1} and x_{i+1} the solutions (1.2) of (2) are valid. The first values of n are $n = 5, 34, 203, 1188, 6929, \ldots$.

For (1.3) the substitution $x = y - 1 = z - 2$ gives $\binom{n}{x} + \binom{n}{x+1} = \binom{n}{x+2}$. The term on the left equals $\binom{n+1}{x+1}$, and all solutions of $\binom{n+1}{x+1} = \binom{n}{x+2}$ have already been determined in [3] to be those of (1.3). The first values of n are $n = 14, 103, 713, 4894, 33551, \ldots$.

It can be remarked that no other solutions than those of Theorem 1 occur for rows $n \leq 200$.

Theorem 2: Equation (3) has at least the following solutions ($n \geq 2$):

$$(x, y, z) = \left(x, \frac{1}{c}\binom{x}{2} - \frac{c-1}{2}, \frac{1}{c}\binom{x}{2} + \frac{c+1}{2}\right), \qquad (2.1)$$

where $c = 1, 2, \ldots$, and x is chosen so that y is an integer (these are all solutions for $n = 2$),

$$(x, y, z) = (2i + 3, 2i + 3, 2i + 4) \text{ for } n = i + 2 \quad (i = 1, 2, \ldots), \qquad (2.2)$$

$$(x, y, z) = (x, x, x + 2) \text{ for } n = n_i, \ x = x_i \ (i = 1, 2, \ldots) \qquad (2.3)$$

with $n_1 = 6$, $x_1 = 19$, $n_{i+1} = 2x_i - n_i + 3$, $x_{i+1} = 7x_i - 4n_i + 9$,

$$(x, y, z) = (x, x + 1, x + 2) \text{ for } n = F_{2i}F_{2i+3} + 1, \text{ and} \qquad (2.4)$$

$$x = F_{2i+2}F_{2i+3} - 1 \quad (i = 1, 2, \ldots).$$

Proof: Equation (3) for $n = 2$ and $z = y + c$ is equivalent to $\binom{x}{2} = cy + \binom{c}{2}$, and (2.1) follows.

The substitution $x = y = z - 1$ in (3) yields $2\binom{x}{n} = \binom{x+1}{n}$ which is equivalent to $\binom{x}{n} = \binom{x}{n-1}$. This, however, is possible only for odd x and $n = \frac{x+1}{2}$, and this is case (2.2).

For (2.3) the substitution $x = y = z - 2$ in (3) implies $2\binom{x}{n} = \binom{x+2}{n}$, and $2(x-n+1)(x-n-2) = (x+1)(x+2)$. This last equation corresponds to the case (1.2) with $x - n$ instead of x, and $2x - n + 2$ instead of n. Thus substitution of x_i and n_i in (1.2) by $x_i - n_i$ and $2x_i - n_i + 2$, respectively, yield

$$2x_{i+1} - n_{i+1} + 2 = 5(2x_i - n_i + 2) + 2(x_i - n_i) + 7,$$

$$x_{i+1} - n_{i+1} = 2(2x_i - n_i + 2) + x_i - n_i + 2,$$

and (2.3) follows by elimination of n_{i+1} and x_{i+1}. The first values of n are $n = 6, 35, 204, 1189, 6930, \ldots$.

The substitution $x = y - 1 = z - 2$ in (3) yields $\binom{x}{n} + \binom{x+1}{n} = \binom{x+2}{n}$ in the case (2.4). The term on the right equals $\binom{x+1}{n} + \binom{x+1}{n-1}$, and as in (1.3) all solutions of $\binom{x}{n} = \binom{x+1}{n-1}$ are deduced from [3]. The first values of n are n = 6, 40, 273, 1870, 12816, ...

All solutions of (3) known at present are contained in Theorem 2.

<u>Theorem 3</u>: Equation (4) has at least the following solutions:

The sporadic solutions (n; x, y, z) = (5; 0, 2, 1),   (3.1)
(6; 0, 1, 2), (6; 1, 3, 2), (7; 1, 3, 2), (9; 0, 3, 2), (11; 0, 4, 2),
(13; 0, 2, 5), and (15; 2, 5, 4),

(x, y, z) = (x, x, x + 1) for n = n_i, x = x_i (i = 1, 2, . . .)   (3.2)
with n_1 = 3, x_1 = 0, n_{i+1} - 1 = $19n_i - 24x_i - 2$, $x_{i+1} = 4n_i - 5x_i - 1$.

<u>Proof</u>: The quadruples of (3.1) are easily checked to fulfil (4).

For (3.2) the substitution of x = y = z - 1 in (4) yields $2\binom{n-x}{x} = \binom{n-x-1}{x+1}$. This is equivalent to 2(n-x) (x+1) = (n-2x) (n-2x-1), and because x $\leq$ n - x it follows that

$$x = \tfrac{1}{6}(3n - 2 - \sqrt{3(n + 1)^2 + 1} \,).$$ (3.3)

Again from [2], p. 94, all solutions of $s^2 - 3t^2 = 1$ are

$(s_1, t_1) = (2, 1), \quad (s_{i+1}, t_{i+1}) = (2s_i + 3t_i, \ s_i + 2t_i)$ (3.4)

with i = 1, 2, ... If s_i of (3.4) is used as the value of the square root in (3.3), then only for odd i the values x of (3.3) are integers, since only in these cases $s_i \equiv 1 \pmod 3$. Hence for n = n_i and x = x_i equation (3.3) is fulfilled by n_i and x_i from $s_{2i-1} = 3n_i - 6x_i - 2$ and $t_{2i-1} = n_i + 1$. Together with the recursions of (3.4) it follows that

$s_{2i+1} = 3n_{i+1} - 6x_{i+1} - 2 = 2s_{2i} + 3t_{2i} = 2(2s_{2i-1} + 3t_{2i-1})$
$\quad + 3(s_{2i-1} + 2t_{2i-1}) = 7s_{2i-1} + 12t_{2i-1} = 7(3n_i - 6x_i - 2) + 12(n_i + 1),$
$t_{2i+1} = n_{i+1} + 1 = s_{2i} + 2t_{2i} = 2s_{2i-1} + 3t_{2i-1} + 2(s_{2i-1} + 2t_{2i-1},$

$$= 4s_{2i-1} + 7t_{2i-1} = 4(3n_i - 6x_i - 2) + 7(n_i + 1).$$

Then by elimination of n_{i+1} and x_{i+1} the solutions (3.2) are obtained. The first values of n are n = 3, 55, 779, 10863, 151315, ...
No other solutions than those of Theorem 3 exist for $n \leq 200$.

REFERENCES

[1]   Heath-Brown, D.R. "The first case of Fermat's Last Theorem." *Math. Intelligencer 7, Nr. 4* (1985): pp 40-47, 55.

[2]   Sierpinski, W. "Elementary Theory of Numbers." *Warszawa* 1964.

[3]   Singmaster, D. "Repeated binomial coefficients and Fibonacci numbers." *The Fibonacci Quarterly 13* (1975): pp 295-298.

F. T. Howard

RECURRENCES RELATED TO THE BESSEL FUNCTION

1. INTRODUCTION

For $k = 0, 1, 2, \ldots$ let $J_k(z)$ be the Bessel function of the first kind. Put

$$(z/2)^k/J_k(z) = \sum_{n=0}^{\infty} u_n(k) \frac{(z/2)^{2n}}{n! \, (n+k)!} \, ; \tag{1.1}$$

then if follows that $u_0(k) = (k!)^2$, and for $n > 0$

$$\sum_{r=0}^{n} (-1)^r \binom{n+k}{r+k} \binom{n+k}{r} u_r(k) = 0.$$

In [4] it was proved that if

$$m = c_0 + c_1 p + c_2 p^2 + \ldots \qquad (0 \le c_0 < p - 2k) \tag{1.2}$$

$$(0 \le c_i < p \text{ for } i > 0)$$

then

$$u_m(k) \equiv u_{c_0}(k). \, w_{c_1} w_{c_2} \ldots (\text{mod } p) \tag{1.3}$$

where p is a prime number, $p > 2k$, and $w_n = u_n(0)$. This extended the work of L. Carlitz [2], who proved the same result for $k = 0$.

In [4] more general recurrences of the type

$$\binom{n+k}{k} H_n = \sum_{r=0}^{n} (-1)^r \binom{n+k}{r+k} \binom{n+k}{r} Q_r G_{n-r} \tag{1.4}$$

7

A. N. Philippou et al. (eds.), Applications of Fibonacci Numbers, 7–16.
© 1988 by Kluwer Academic Publishers.

were also studied, where H_n, Q_n and G_n are polynomials with coefficients that are integral (mod p) for a given prime $p > 2k$. The results, which were congruences similar to (1.3), depended on H_0 and Q_0 both being nonzero in some cases.

The purpose of the present paper is to investigate recurrences of the type (1.4) with $H_0 = Q_0 = 0$. The main results, Theorem 1 and Theorem 2 in section 3, give congruences for H_m if m is of the form (1.2) and if G_n and Q_n are given polynomials with certain properties; congruences for Q_m are obtained if G_n and H_n are given. These results are in a somewhat more general setting than the theorems in [2] and [4].

The following application is discussed in section 4. Define $a_n(k)$ by means of

$$a_n(k) = 2^{2n} n! \, (n + k)! \, \sigma_{2n}(k),$$

where $\sigma_{2n}(k)$ is the Rayleigh function [5], [6]. We show that $a_n(k)$ satisfies a recurrence of the type (1.4) and is integral (mod p) if $p > 2k$. Applying Theorem 2, we prove

$$a_n(k) \equiv 0 \ (\text{mod } p^t) \ \text{if } p^t \mid n \ (t = 1, 2, \ldots),$$
$$a_m(k) \equiv 0 \ (\text{mod } p)$$

if p is a prime number, $p > 2k$, and if $m \geq p$ is of the form (1.2). These congruences extend the work of Carlitz [1] and the writer [3], who studied the cases $a_n(0)$ and $a_n(1)$ respectively. By applying Theorem 1 we also obtain congruences for the polynomials $a_n(k; x)$ defined by

$$a_n(k; x) = \sum_{r=0}^{n} (-1)^r \begin{pmatrix} n+k \\ r+k \end{pmatrix} \begin{pmatrix} n+k \\ r \end{pmatrix} a_r(k) \, x^{n-r}.$$

2. PRELIMINARIES

The first two lemmas are due to Lucas [8] and Kummer [7] respectively. We use the notation $p^t \parallel n$ to mean $p^t \mid n$ and $p^{t+1} \nmid n$.

Lemma 1. If p is a prime number and

$$n = n_0 + n_1 p + \ldots + n_j p^j \qquad (0 \le n_i < p),$$
$$r = r_0 + r_1 p + \ldots + r_j p^j \qquad (0 \le r_i < p),$$

then

$$\binom{n}{r} \equiv \binom{n_0}{r_0} \binom{n_1}{r_1} \cdots \binom{n_j}{r_j} \qquad (\text{mod } p).$$

Lemma 2. With the hypothesis of Lemma 1, let

$$n - r = s_0 + s_1 p + \ldots + s_j \, p^j \qquad (0 \le s_i < p),$$

and suppose
$$r_0 + s_0 = e_0 \, p + d_0 \qquad\qquad (0 \le d_0 < p)$$
$$e_0 + r_1 + s_1 = e_1 \, p + d_1 \qquad\qquad (0 \le d_1 < p)$$
$$\ldots$$
$$e_{j-1} + r_j + s_j = e_j \, p + d_j \qquad\qquad (0 \le d_j < p).$$

then $p^M \parallel \binom{n}{r}$, where $M = e_0 + e_1 + \ldots + e_j$.

It follows from Lemma 1 that if p is a prime number, then

$$\binom{np}{rp} \equiv \binom{n}{r} \qquad (\text{mod } p).$$

Also by Lemma 1, if $p - 2k > s \ge 0$, then for $j = s + 1, s + 2, \ldots, p - 1$,

$$\binom{np+s+k}{rp+j+k} \binom{np+s+k}{rp+j} \equiv 0 \qquad (\text{mod } p).$$

It follows from Lemma 2 that if $r_i > n_i$ and $r_{i+v} \ge n_{i+v}$ for $v = 1, \ldots, t - 1$, then

$$\binom{n}{r} \equiv 0 \qquad (\text{mod } p^t).$$

The following lemma was proved in [4].

Lemma 3. Let p be a prime number, $p > 2k$.

Then $\begin{bmatrix} n+k \\ r+k \end{bmatrix} \begin{bmatrix} n+k \\ r \end{bmatrix} / \begin{bmatrix} n+k \\ k \end{bmatrix}$

is integral (mod p) for $r = 0, 1, \ldots, n$.

3. MAIN RESULTS

In the following two theorems we make these assumptions: Let k be a fixed nonnegative integer, let p be a prime number, $p > 2k$, and let $\{F_n\}$, $\{G_n\}$ and $\{R_n\}$ be polynomials in an arbitrary number of indeterminates with coefficients that are integral (mod p).

Assume

$$G_0 = R_0 = 1,$$
$$G_m \equiv G_{c_0} R_{c_1}^p R_{c_2}^{p^2} \ldots \text{ (mod p)}$$

for any m of the form (1.2); also assume

$$F_0 = 0,$$
$$F_n \equiv 0 \text{ (mod } p^t) \text{ if } p^t \mid n \text{ } (t = 1, 2, \ldots).$$

Theorem 1: Suppose

$$F_m \equiv 0 \text{ (mod p)} \tag{3.1}$$

for any $m \geq p$ of the form (1.2). Define $\{H_n\}$ by

$$\begin{bmatrix} n+k \\ k \end{bmatrix} H_n = \sum_{r=0}^{n} (-1)^r \begin{bmatrix} n+k \\ r+k \end{bmatrix} \begin{bmatrix} n+k \\ r \end{bmatrix} F_r G_{n-r}. \tag{3.2}$$

Then the polynomials H_n have coefficients that are integral (mod p) for $p > 2k$, $H_0 = 0$, and

$$H_n \equiv 0 \text{ (mod } p^t) \text{ if } p^t \mid n \qquad (t = 1, 2, \ldots), \tag{3.3}$$

$$H_m \equiv H_{c_0} R_{c_1}^p R_{c_2}^{p^2} \ldots \text{ (mod p)}$$

for any $m \geq p$ of the form (1.2).

Proof. The coefficients of H_n are integral (mod p) by Lemma 3. Also, since $F_0 = 0$ it follows immediately from (3.2) that $H_0 = 0$. Now suppose $p^t \parallel n$. We will show that p^t divides each term on the right side of (3.2). For each r, $0 \leq r \leq n$, we have one of the following cases.

Case 1: $p^s \mid r$, $s \geq t$. Then $p^t \mid F_r$.

Case 2: $p^s \parallel r$, $0 < s < t$. Then $p^s \mid F_r$ and $p^{t-s} \mid \binom{n+k}{r+k}$.

Case 3: $p \nmid r$. Then $\binom{n+k}{r+k} \equiv 0 \pmod{p^t}$ $(r + k < p)$.

Suppose $r + k = hp + j$, $0 \leq j < p$. Then

$$\binom{n+k}{r+k} \equiv 0 \pmod{p^t} \ (j > k).$$

If $j \leq k$, then $r = (h-1)p + (p - k + j)$. Since $p - k + j > k$, we have

$$\binom{n+k}{r} \equiv 0 \pmod{p^t}.$$

Thus

$$H_n \equiv 0 \pmod{p^t}.$$

Now suppose m is of the form (1.2). Then by (3.1), (3.2) and Lemma 1, we have

$$\binom{c_0+k}{k} H_m \equiv \sum_{r=0}^{c_0} (-1)^r \binom{c_0+k}{r+k} \binom{c_0+k}{r} F_r G_{c_0-r} R_{c_1}^p \ R_{c_2}^{p^2} \ \ldots$$

$$\equiv \binom{c_0+k}{k} H_{c_0} \ R_{c_1}^p \ R_{c_2}^{p^2} \ \ldots \pmod{p},$$

and the proof is complete.

Theorem 2: Suppose

$$F_m \equiv F_{c_0} \ R_{c_1}^p \ R_{c_2}^{p^2} \ \ldots \pmod{p} \tag{3.4}$$

for any m of the form (1.2). Define $\{Q_n\}$ by

$$\binom{n+k}{k} F_n = \sum_{r=0}^{n} (-1)^r \binom{n+k}{r+k} \binom{n+k}{r} Q_r G_{n-r}. \tag{3.5}$$

Then each Q_n has coefficients that are integral (mod p) for $p > 2k$, $Q_0 = 0$, and

$$Q_n \equiv 0 \ (\text{mod } p^t) \ \text{if} \ p^t \mid n \qquad (t = 1, 2, \ldots) \tag{3.6}$$

$$Q_m \equiv 0 \ (\text{mod } p) \tag{3.7}$$

for any $m \geq p$ of the form (1.2).

Proof: We first rewrite (3.5) as

$$(-1)^n \binom{n+k}{k} Q_n = \binom{n+k}{k} F_n + \sum_{r=0}^{n-1} (-1)^{r-1} \binom{n+k}{r+k} \binom{n+k}{r} Q_r G_{n-r}. \tag{3.8}$$

Since $F_0 = 0$, we see that $Q_0 = 0$; also, by using induction on n, along with Lemma 3, we see that Q_n has coefficients that are integral (mod p). We now prove (3.6) by induction on n. Congruence (3.6) is true for $n = 0$; assume it is true for $n = 0, \ldots, N-1$. After replacing n by N in (3.8), we see that the proof is now identical to the proof of (3.3) in Theorem 1.

To prove (3.7), we first observe that by (3.8) and Lemma 1,

$$Q_p \equiv 0 \ (\text{mod } p) \qquad (p > 2k).$$

Now assume that m is of the form (1.2) and assume that

$$Q_n \equiv 0 \ (\text{mod } p)$$

For all $n < m$ such that $n = hp + j$, $h \geq 1$, $0 \leq j < p - 2k$. By (3.8) and Lemma 1, we have

$$(-1)^m \binom{c_0+k}{k} Q_m$$

$$\equiv \left\{ \binom{c_0+k}{k} F_{c_0} + \sum_{r=0}^{c_0} (-1)^{r-1} \binom{c_0+k}{r+k} \binom{c_0+k}{r} Q_r G_{c_0-r} \right\} R_{c_1}^p R_{c_2}^{p^2} \cdots$$

$$\equiv 0 \pmod p.$$

This completes the proof of Theorem 2.

4. APPLICATIONS

We first observe that if

$$B_k(z) = \sum_{n=0}^{\infty} b_n(k) \frac{(z/2)^{2n}}{n! \, (n+k)!}$$

and

$$D_k(z) = \sum_{n=0}^{\infty} (-1)^n \, d_n(k) \frac{(z/2)^{2n}}{n! \, (n+k)!}$$

are formal generating functions, and if

$$B_k(z) = (z/2)^k \, D_k(z)/J_k(z),$$

then

$$\binom{n+k}{k} k! \, d_n(k) = \sum_{r=0}^{n} (-1)^r \binom{n+k}{r+k} \binom{n+k}{r} b_r(k).$$

Example 1: For $k = 0, 1, 2, \ldots$ define $a_n(k)$ by

$$a_n(k) = 2^{2n} n! \, (n+k)! \, \sigma_{2n}(k),$$

where $\sigma_{2n}(k)$ is the Rayleigh function [5], [6]. The integers $a_n(0)/n$ and $a_n(1)$ have been studied in some detail by L. Carlitz [1] and the writer [3]. A generating function is

$$(z/2)J_{k+1}(z)/J_k(z) = \sum_{n=0}^{\infty} a_n(k) \frac{(z/2)^{2n}}{(n+k)! \, n!} \tag{4.1}$$

where $J_k(z)$ is the Bessel function of the first kind; it follows that

$$\binom{n+k}{k} (-n \cdot k!) = \sum_{r=0}^{n} (-1)^r \binom{n+k}{r+k} \binom{n+k}{r} a_r(k).$$

By Theorem 2, $a_n(k)$ is integral (mod p) for any prime $p > 2k$, $a_0(k) = 0$, and

$$a_n(k) \equiv 0 \pmod{p^t} \text{ if } p^t \mid n \quad (n = 1, 2, \ldots),$$

$$a_m \equiv 0 \pmod{p}$$

if $m \geq p$ is of the form (1.2).

Next define the polynomial $a_n(k; x)$ by

$$a_n(k; x) = \sum_{r=0}^{n} (-1)^r \begin{bmatrix} n+k \\ r+k \end{bmatrix} \begin{pmatrix} n+k \\ r \end{pmatrix} a_r(k) \, x^{n-r}.$$

It follows that $a_0(k; x) = 0$ and that $a_n(k; x)$ has coefficients that are integral (mod p) for $p > 2k$. By Theorem 1,

$$a_n(k; x) \equiv 0 \pmod{p^t} \text{ if } p^t \mid n \quad (t = 1, 2, \ldots),$$

$$a_m(k; x) \equiv a_{c_0} (k; x) \, x^{m-c_0} \pmod{p}$$

if $m > p$ is of the form (1.2). A generating function is

$$(z/2)^{1-k} x^{-k/2} J_{k+1}(z) J_k(z\sqrt{x}) \, / \, J_k(z)$$

$$= \sum_{n=0}^{\infty} (-1)^n \, a_n(k; x) \, \frac{(z/2)^{2n}}{(n+k)!(n+k)!}$$

It might be of interest to note the relationship of $a_n(k)$ to the Rayleigh polynomial $\varnothing_{2n}(k)$ defined by Kishore in [6]. It is

$$a_n(k) = n! k! \varnothing_{2n}(k) \, / \, \prod_{j=1}^{n} (k+j)^{[n/j]-1}.$$

It is also easy to find a relationship to the numbers $u_n(k)$ defined by (1.1). By (1.1) and (4.1) we have

$$\begin{pmatrix} n+k \\ k \end{pmatrix} (-1)^n k! \, a_n(k) = \sum_{r=0}^{n-1} (-1)^{r-1} \begin{bmatrix} n+k \\ r+k \end{bmatrix} \begin{pmatrix} n+k \\ r \end{pmatrix} (n-r) u_r(k).$$

Because of properties of the Bessel function like

$$zJ_{k+2}(z) = 2(k+1)J_{k+1}(z) - zJ_k(z),$$

t is possible to write down generating functions for $a_n(k)$ in several different ways. For example,

$$(z^2/4)J_{k+2}(z)/J_k(z) = (k+1) \sum_{n=2}^{\infty} a_n(k) \frac{(z/2)^{2n}}{(n+k)! \; n!},$$

$$(z/2)J_{k-1}(z)/J_k(z) = k - \sum_{n=1}^{\infty} a_n(k) \frac{(z/2)^{2n}}{(n+k)!n!},$$

$$\{J_{k+1}(z)/J_k(z)\}^2 = \sum_{n=1}^{\infty} a_{n+1}(k) \frac{(z/2)^{2n}}{(n+k)!(n+1)!},$$

and so on.

Example 2: For $0 \le s \le k + 1$, define $a_n^{(s)}(k)$ by means of

$$(z/2)^{k+2-s} J_s(z)/J_k(z) = \sum_{n=0}^{\infty} a_n^{(s)}(k) \frac{(z/2)^{2n}}{(n+k)!n!}.$$

Then we have

$$\binom{n+k}{k} \{-n \cdot k! \, (n + s) \, (n + s + 1) \ldots (n + k) \}$$

$$= \sum_{r=0}^{n} (-1)^r \binom{n+k}{r+k} \binom{n+k}{r} a_r^{(s)}(k).$$

We can apply Theorem 2: The numbers $a_r^{(s)}(k)$ are integral (mod p) for $p > 2k$,

$$a_0^{(s)}(k) = 0, \text{ and}$$
$$a_n^{(s)}(k) \equiv 0 \pmod{p^t} \text{ if } p^t \mid n \quad (t = 1, 2, \ldots),$$
$$a_m^{(s)}(k) \equiv 0 \pmod{p}$$

if $m \ge p$ is of the form (1.2).

If we define the polynomial $a_n^{(s)}(k; x)$ by

$$a_n^{(s)}(k; x) = \sum_{r=0}^{n} (-1)^r \binom{n+k}{r+k} \binom{n+k}{r} a_r^{(s)}(k)x^{n-r},$$

then by Theorem 1 we have

$$a_n^{(s)}(k; x) \equiv 0 \pmod{p^t} \text{ if } p^t \mid n \ (t = 1, 2, \ldots),$$
$$a_m^{(s)}(k; x) \equiv a_{c_0}^{(s)}(k; x)x^{m-c_0} \pmod{p}$$

if $m \geq p$ is of the form (1.2). A generating function is

$$(z/2)^{2-s}x^{-k/2}J_s(z) \ J_k \ (z \ \sqrt{x})/J_k(z) = \sum_{n=0}^{\infty} (-1)^n \ a_n^{(s)}(k; x) \ \frac{(z/2)^{2n}}{(n+k)!(n+k)!} \ .$$

We observe that

$$a_n^{(k+1)}(k) - a_n(k),$$
$$a_n^{(k-1)}(k) = -n(n+k) \ a_{n-1}(k) \qquad (n > 1).$$

REFERENCES

[1] Carlitz, L. "A Sequence of Integers Related to the Bessel Functions." *Proc. Amer. Math. Soc. 14* (1963): pp 1-9.

[2] Carlitz, L. "The Coefficients of the Reciprocal of $J_0(x)$." *Arc. Math. 6* (1955): pp 121-127.

[3] Howard, F. T. "Integers Related to the Bessel Function $J_1(z)$." *The Fibonacci Quarterly 23* (1985): pp 249-257.

[4] Howard, F. T. "The Reciprocal of the Bessel Function $J_k(z)$." To appear.

[5] Kishore, N. "The Rayleigh Function." *Proc. Amer. Math. Soc. 14* (1963): pp 527-533.

[6] Kishore, N. "The Rayleigh Polynomial." *Proc. Amer. Math. Soc. 15* (1964): pp 911-917.

[7] Kummer, E. "Uber die Erganzungssatze zu den Allegemeinen Reciprocitatsgesetzen." *J. Reine Angeu. Math. 44* (1852): pp 93-146.

[8] Lucas, E. "Sur les congruences des nombres euleriens et des coefficients differentiels . . . " *Bull. Soc. Math. France 6* (1878): pp 49-54.

Kenji Nagasaka and Shiro Ando

SYMMETRIC RECURSIVE SEQUENCES MOD M

Distribution properties of integer sequences have been widely studied from various points of view. The sequence of Fibonacci numbers $\{F_n\}$ is, of course, one of the main targets for this study. Indeed, $\{\log F_n\}$ is uniformly distributed mod 1, so that $\{F_n\}$ obeys Benford's law, detailed study of which is carried out in [6]. In his note we are going to treat uniform distribution properties of certain recursive integer sequences in residue classes.

Uniformly distributed integer sequences in residue classes have already been considered since the beginning of this century, when L. E. Dickson [2] studied permutation polynomials, i.e. polynomials inducing a permutation of residue classes with respect to a fixed prime number. Independently of the study of permutation polynomials, I. Niven [12] introduced the notion of uniformly distributed sequences mod m from a general point of view as follows:

Definition 1: An integer sequence a = $\{a_n\}$ is called to be uniformly distributed mod m, if, for every j,

$$\lim_{N \to \infty} \frac{A_N(j, m; a)}{N} = \frac{1}{m}, \quad j = 0, 1, \ldots, m - 1, \tag{1}$$

where m is a fixed positive integer greater than one and $A_N(j, m; a)$ denotes the number of indices n up to N satisfying the following congruence:

$$a_N \equiv j \pmod{m}. \tag{2}$$

This notion is a particular case of the concept of uniform distribution in compact abelian groups, but the general approach is usually of no help. The

17

A. N. Philippou et al. (eds.), Applications of Fibonacci Numbers, 17–28.

sequence of Fibonacci numbers is uniformly distributed mod m only for $m = 5^k$ (k = 1, 2, . . .), proved by Lauwerence Kuipers and Jau-Shyong Shiue [3], [4] and Harald Niederreiter [11]. Furthermore, the problem of characterizing integers m for which a given second-order recurrence is uniformly distributed mod m was tackled and solved almost completely by R. T. Bumby [1].

Since such characterized integers m are rather limited, we are led to relax the Definition 1 and consider two definitions as follows:

Definition 2: A sequence of integers a = {a_n} is weakly uniformly distributed mod m, if for every j relatively prime to m,

$$\lim_{N \to \infty} \frac{A_N(j, m ; a)}{B_N(m ; a)} = \frac{1}{\varnothing(m)} ,$$

(3

provided that there exist infinitely many n for which a_n is relatively prime to m. $\varnothing(\cdot)$ is the Euler's tortient function and $A_N(j, m; a)$ is the same notation as in Definition 1 and $B_N(m; a)$ denotes the number of indices n between 1 and N such that a_n is relatively prime to m.

This definition is due to Narkiewicz [9].

Definition 3:. A sequence of integers a = {a_n} is uniformly distributed in $(\mathbb{Z}/m\mathbb{Z})^*$, if, for every invertible element j in $\mathbb{Z} / m\mathbb{Z}$,

$$\lim_{N \to \infty} \frac{A_N(j , m; a)}{N} = \frac{1}{\varnothing(m)} ,$$

(4

if every a_n is relatively prime to the modulus m. Note that $(\mathbb{Z}/m\mathbb{Z})^*$ is the multiplicative group of invertible elements j relatively prime to m.

This definition was first introduced by Nagasaka [5]. Clearly, an integer sequence uniformly distributed in $(\mathbb{Z}/m\mathbb{Z})^*$ is weakly uniformly distributed mod m but the converse is not always true.

The standard example here is the sequence of prime numbers, which is weakly uniformly distributed mod m for every modulus m by Dirichlet's prime number theorem in arithmetic progressions. K. Nagasaka [5] studied the uniform distribution property in $(Z/mZ)^*$ for special recursive sequences which satisfy either

$$u_{n+1} \equiv u_n + u_n^{-1} \pmod{m} \tag{5}$$

or

$$u_{n+1} \equiv a \cdot u_n + b \cdot u_n^{-1} \pmod{m}, \tag{6}$$

where a and b are integers relatively prime to m.

These recurrence formulae satisfy a functional equation when we consider these recurrence formulae as a mapping f from $(Z/mZ)^*$ to $(Z/mZ)^*$.

In [8] Nagasaka extended results obtained in [5] and treated recursive sequences satisfying a symmetric functional equation:

$$f(s) = f(s^{-1}) \tag{7}$$

on $(Z/mZ)^*$ and also other recursive sequences satisfying

$$w_{n+1} \equiv a \cdot w_n^k + b \cdot w_n^{-k} \pmod{m}, \tag{8}$$

where a and b are invertible elements in $(Z/mZ)^*$.

First, we want to remark on the invertibility of a and b in (6) and (8). The proof of Theorem 4 in [5] is based on the functional equation:

$$f(s) = f(a^{-1} \cdot b \cdot s^{-1}) \tag{9}$$

for all s in $(Z/mZ)^*$. Hence we need the invertibility.

Without using the functional equation (9), a necessary condition for the recursive sequence $\{u_n\}$ satisfying (5) to be uniformly distributed in $(Z/mZ)^*$ is given by Lemma 1 [7]. By applying this Lemma 1 for the recursive sequence $\{u_n\}$ satisfying (6), if $\{u_n\}$ is uniformly distributed in $(Z/mZ)^*$, then

$$(a^2 + b^2 - 1) \cdot \sum_{(i,m)=1} i^2 + 2 \cdot ab \cdot \emptyset(m) \equiv 0 \pmod{m}. \tag{10}$$

For example, setting $a = 2$, $b = 3$ and $m = 6$, the (10) is valid, and $\{u_n\}$ satisfying the recurrence formula:

$$u_{n+1} \equiv 2 \cdot u_n + 3 \cdot u_n^{-1} \pmod{6} \tag{11}$$

is uniformly distributed in $(\mathbb{Z}/6\mathbb{Z})^*$.

This remark is also valid for the sequence $\{w_n\}$ generated by (8). Indeed, the sequence $\{w_n\}$ satisfying

$$w_{n+1} \equiv 2 \cdot w_n^3 + 3 \cdot w_n^{-3} \pmod{6} \tag{12}$$

is uniformly distributed in $(\mathbb{Z}/6\mathbb{Z})^*$.

Let us consider symmetric recursive sequences mod m. We call an integer sequence $t = \{t_n\}$ a symmetric recursive sequence mod m if $\{t_n\}$ is generated by a recurrence formula of order 1:

$$t_{n+1} \equiv f(t_n) \pmod{m}, \tag{13}$$

where f is a polynomial of s and s^{-1} and satisfies the functional equation (7) on $(\mathbb{Z}/m\mathbb{Z})^*$. In [8] we considered symmetric recursive sequences $\{v_n\}$ generated by

$$\begin{aligned}
v_{n+1} \equiv a_k \, (v_n^k + v_n^{-k}) + a_{k-1} \, (v_n^{k-1} + v_n^{-(k-1)}) + \ldots \\
+ a_1 \, (v_n + v_n^{-1}) + a_0 \pmod{m}.
\end{aligned} \tag{14}$$

The corresponding mapping g to (14) from $(\mathbb{Z}/m\mathbb{Z})^*$ to $(\mathbb{Z}/m\mathbb{Z})^*$ satisfies (7) obviously, but there exist symmetric recursive sequences other than (14). Henceforth, we are going to treat general symmetric recursive sequences mod m which are uniformly distributed in $(\mathbb{Z}/m\mathbb{Z})^*$. Then, we have necessarily, for every s in $(\mathbb{Z}/m\mathbb{Z})^*$,

$$s \equiv s^{-1} \pmod{m}$$

or equivalently

$$s^2 \equiv 1 \ (\text{mod } m). \tag{15}$$

The modulus m is represented in the form:

$$m = p_1^{e_1} \cdot p_2^{e_2} \ldots p_r^{e_r},$$

where $p_1 < p_2 < \ldots < p_r$ are prime numbers and the e_i's are positive integers. Then, from (15)

$$s^2 \equiv 1 \ (\text{mod } p_i^{e_i}), \ i = 1, 2, \ldots, r, \tag{16}$$

for every $s \in (Z/p_i^{e_i} Z)^*$. Suppose further that $p_1 \geq 5$. Then $3 \in (Z/mZ)^*$. But $3^2 = 9$ cannot be congruent to 1 for any modulus $p_i^{e_i}$ with $p_i \geq 5$. Hence we may restrict ourselves to the case in which either $m = 2^\alpha$ or $m = 3^\beta$. Then by substituting an invertible element 3 in (15), we see that (15) is satisfied for every s in $(Z/2^\alpha Z)^*$ only when α is less than 4. Similarly by considering an invertible element 2, (15) is valid for every s in $(Z/3^\beta Z)^*$ only if $\beta = 1$. Summarizing the argument above, we have

Theorem 1: Let $\{t_n\}$ be a symmetric recursive sequence which is defined by the congruence (13) and satisfies (7) . If the sequence $\{t_n\}$ is uniformly distributed in $(Z/mZ)^*$, then m is necessarily of the form $2^\alpha \cdot 3^\beta$ for α = 0, 1, 2, 3, and β = 0, 1 .

Now we are going to settle the question of determining all symmetric recursive sequences, which are uniformly distributed in $(Z/mZ)^*$. From Theorem 1, we may suppose that the modulus m is one of the following: 2, 3, 4, 6, 8, 12, and 24. It is enough to consider the case that the recurrence f in (13) is a polynomial only of s. Hence since every invertible element in $(Z/mZ)^*$ for the above m is of order 2, we may suppose that

$$f(s) = a \cdot s + b ,$$

where a and b are elements of $\mathbb{Z}/m\mathbb{Z}$.

Since we employ the Definition 3, we agree to say that a symmetric recursive sequence $\{t_n\}$ is uniformly distributed in $(\mathbb{Z}/m\mathbb{Z})^*$ if $\{t_n\}$ is uniformly distributed in $(\mathbb{Z}/m\mathbb{Z})^*$ for every initial value $t_1 \in (\mathbb{Z}/m\mathbb{Z})^*$. Thus we generate all symmetric recursive sequences $\{t_n\}$ by setting their initial values t_1 to be 1 and checking whether $\{t_n\}$ is uniformly distributed in $(\mathbb{Z}/m\mathbb{Z})^*$ for every a and every b in $\mathbb{Z}/m\mathbb{Z}$ with $m \in \{2, 3, 4, 6, 8, 12, 24\}$. Then we obtain:

Theorem 2: Let $\{t_n\}$ be a symmetric recursive sequence mod m which is generated by the congruence (13) and satisfies (7). $\{t_n\}$ is uniformly distributed in $(\mathbb{Z}/m\mathbb{Z})^*$, only when the recurrence formula of $\{t_n\}$ is one of the following:

$$t_{n+1} \equiv t_n \pmod 2.$$
$$t_{n+1} \equiv 2 \cdot t_n \pmod 3.$$
$$t_{n+1} \equiv t_n + 2 \pmod 4.$$
$$t_{n+1} \equiv 3 \cdot t_n \pmod 4.$$
$$t_{n+1} \equiv 2 \cdot t_n + 3 \pmod 6.$$
$$t_{n+1} \equiv 5 \cdot t_n \pmod 6.$$
$$t_{n+1} \equiv t_n + 2 \pmod 8.$$
$$t_{n+1} \equiv t_n + 6 \pmod 8.$$
$$t_{n+1} \equiv 5 \cdot t_n + 2 \pmod 8.$$
$$t_{n+1} \equiv 5 \cdot t_n + 6 \pmod 8.$$

Remark 1: The representation of a recurrence formula is not unique. For example, the following formula:

$$t_{n+1} \equiv 2 \cdot t_n \pmod 3$$

is identical to

$$t_{n+1} \equiv t_n + t_n^{-1} \pmod 3,$$

which appeared in [5], [7] and [8].

Remark 2: If we omit the assumption that f is a polynomial of s and s^{-1}, there exist other recursive sequences satisfying (13) which are uniformly distributed in $(Z/mZ)^*$ since every cyclic purmutation over $(Z/mZ)^*$ gives a uniformly distributed sequence $\{c_n\}$ on $(Z/mZ)^*$. This sequence is not always to be represented by (13) with a polynomial f. Further, there exist $\{\emptyset(m) - 1\}!$ recurrences f in (13) by which the generated sequences are uniformly distributed in $(Z/mZ)^*$.

Hence for the cases of moduli 4 and 6, two recurrences in Theorem 2 generate an identical sequence. For the case of modulus 8, the recurrences $t_{n+1} \equiv t_n + 2 \pmod 8$, and

$$t_{n+1} \equiv 5 \cdot t_n + 6 \pmod 8$$

in Theorem 2 generate an identical sequence and the other two recurrences mod 8 in the above Theorem do also.

Now let us consider weakly uniform distributions mod m. Symmetric recursive sequence $t = \{t_n\}$ mod m are defined by (13) and f satisfies the functional equation (7) on $(Z/mZ)^*$.

Thus, similarly to the case of uniform distribution in $(Z/mZ)^*$, we have:

Theorem 3: Let $t = \{t_n\}$ be a symmetric recursive sequence which is defined by the congruence (13) and satisfies (7). If the sequence $\{t_n\}$ is weakly uniformly distributed mod m , then m is necessarily of the form $2^\alpha \cdot 3^\beta$ for $\alpha = 0, 1, 2, 3$ and $\beta = 0, 1$.

The functional equation (7) is meaningless for a non-invertible element s, since s^{-1} cannot be defined. Hence we can not assume the congruence (15) for a non-invertible element s. Thus we consider integer sequences $t = \{t_n\}$ mod m for the above m in Theorem 3 which satisfy a linear recurence congruence of order 1:

$$t_{n+1} \equiv a \cdot t_n + b \pmod m, \tag{17}$$

with $a = 1, 2, \ldots, m - 1$ and $b \in \mathbf{Z}/m\mathbf{Z}$. Then we get:

<u>Theorem 4</u>: <u>Let</u> $\{t_n\}$ <u>be a</u> <u>symmetric</u> <u>recursive</u> <u>sequence</u> <u>mod</u> m <u>which</u> <u>is</u> <u>defined</u> <u>by</u> (17) <u>and</u> <u>for</u> <u>which</u> m <u>is</u> <u>of</u> <u>the</u> <u>form</u> <u>in</u> <u>Theorem</u> <u>3</u>. <u>The</u> <u>sequence</u> $\{t_n\}$ <u>is</u> <u>weakly</u> <u>uniformly</u> <u>distributed</u> <u>mod</u> m, <u>only</u> <u>when</u> <u>the</u> <u>recurrence</u> <u>formula</u> <u>of</u> $\{t_n\}$ <u>is</u> <u>one</u> <u>of</u> <u>the</u> <u>following</u>. <u>The</u> <u>symbol</u> (∗) <u>after</u> <u>the</u> <u>recurrence</u> <u>formula</u> <u>signifies</u> <u>that</u> $\{t_n\}$ <u>is</u> <u>also</u> <u>uniformly</u> <u>distributed</u> <u>mod</u> m.

(i) m = 2 .

$$t_{n+1} \equiv t_n \ (\text{mod } 2).$$
$$t_{n+1} \equiv t_n + 1 \ (\text{mod } 2) \ (∗)$$

(ii) m = 3 .

$$t_{n+1} \equiv t_n + 1 \ (\text{mod } 3) \ (∗).$$
$$t_{n+1} \equiv t_n + 2 \ (\text{mod } 3) \ (∗).$$
$$t_{n+1} \equiv 2 \cdot t_n \ (\text{mod } 3).$$

(iii) m = 4 .

$$t_{n+1} \equiv t_n + 1 \ (\text{mod } 4) \ (∗).$$
$$t_{n+1} \equiv t_n + 2 \ (\text{mod } 4).$$
$$t_{n+1} \equiv t_n + 3 \ (\text{mod } 4) \ (∗).$$
$$t_{n+1} \equiv 3 \cdot t_n \ (\text{mod } 4).$$

(iv) m = 6 .

$$t_{n+1} \equiv t_n + 1 \ (\text{mod } 6) \ (∗).$$
$$t_{n+1} \equiv t_n + 2 \ (\text{mod } 6).$$
$$t_{n+1} \equiv t_n + 4 \ (\text{mod } 6).$$
$$t_{n+1} \equiv t_n + 5 \ (\text{mod } 6) \ (∗).$$
$$t_{n+1} \equiv 2 \cdot t_n + 3 \ (\text{mod } 6).$$
$$t_{n+1} \equiv 4 \cdot t_n + 1 \ (\text{mod } 6).$$
$$t_{n+1} \equiv 4 \cdot t_n + 5 \ (\text{mod } 6).$$
$$t_{n+1} \equiv 5 \cdot t_n \ (\text{mod } 6).$$

(v) m = 8 .

$$t_{n+1} \equiv t_n + 1 \ (\text{mod } 8) \ (*).$$
$$t_{n+1} \equiv t_n + 2 \ (\text{mod } 8).$$
$$t_{n+1} \equiv t_n + 3 \ (\text{mod } 8) \ (*).$$
$$t_{n+1} \equiv t_n + 5 \ (\text{mod } 8) \ (*).$$
$$t_{n+1} \equiv t_n + 6 \ (\text{mod } 8).$$
$$t_{n+1} \equiv t_n + 7 \ (\text{mod } 8) \ (*).$$
$$t_{n+1} \equiv 5 \cdot t_n + 1 \ (\text{mod } 8) \ (*).$$
$$t_{n+1} \equiv 5 \cdot t_n + 2 \ (\text{mod } 8).$$
$$t_{n+1} \equiv 5 \cdot t_n + 3 \ (\text{mod } 8) \ (*).$$
$$t_{n+1} \equiv 5 \cdot t_n + 5 \ (\text{mod } 8) \ (*).$$
$$t_{n+1} \equiv 5 \cdot t_n + 6 \ (\text{mod } 8).$$
$$t_{n+1} \equiv 5 \cdot t_n + 7 \ (\text{mod } 8) \ (*).$$

(vi) m = 12 .

$$t_{n+1} \equiv t_n + 1 \ (\text{mod } 12) \ (*).$$
$$t_{n+1} \equiv t_n + 2 \ (\text{mod } 12).$$
$$t_{n+1} \equiv t_n + 5 \ (\text{mod } 12) \ (*).$$
$$t_{n+1} \equiv t_n + 7 \ (\text{mod } 12) \ (*).$$
$$t_{n+1} \equiv t_n + 10 \ (\text{mod } 12).$$
$$t_{n+1} \equiv t_n + 11 \ (\text{mod } 12) \ (*).$$
$$t_{n+1} \equiv 7 \cdot t_n + 4 \ (\text{mod } 12).$$
$$t_{n+1} \equiv 7 \cdot t_n + 8 \ (\text{mod } 12).$$

(vii) m = 24.

$$t_{n+1} \equiv t_n + 1 \ (\text{mod } 24) \ (*).$$
$$t_{n+1} \equiv t_n + 2 \ (\text{mod } 24).$$
$$t_{n+1} \equiv t_n + 5 \ (\text{mod } 24) \ (*).$$
$$t_{n+1} \equiv t_n + 7 \ (\text{mod } 24) \ (*).$$
$$t_{n+1} \equiv t_n + 11 \ (\text{mod } 24) \ (*).$$
$$t_{n+1} \equiv t_n + 13 \ (\text{mod } 24) \ (*).$$
$$t_{n+1} \equiv t_n + 14 \ (\text{mod } 24).$$
$$t_{n+1} \equiv t_n + 17 \ (\text{mod } 24) \ (*).$$
$$t_{n+1} \equiv t_n + 19 \ (\text{mod } 24) \ (*).$$

$$t_{n+1} \equiv t_n + 22 \pmod{24}.$$

$$t_{n+1} \equiv t_n + 23 \pmod{24} (*).$$

$$t_{n+1} \equiv 13 \cdot t_n + 1 \pmod{24} (*).$$

$$t_{n+1} \equiv 13 \cdot t_n + 2 \pmod{24}.$$

$$t_{n+1} \equiv 13 \cdot t_n + 5 \pmod{24} (*).$$

$$t_{n+1} \equiv 13 \cdot t_n + 7 \pmod{24} (*).$$

$$t_{n+1} \equiv 13 \cdot t_n + 10 \pmod{24}.$$

$$t_{n+1} \equiv 13 \cdot t_n + 11 \pmod{24} (*).$$

$$t_{n+1} \equiv 13 \cdot t_n + 13 \pmod{24} (*).$$

$$t_{n+1} \equiv 13 \cdot t_n + 14 \pmod{24}.$$

$$t_{n+1} \equiv 13 \cdot t_n + 17 \pmod{24} (*).$$

$$t_{n+1} \equiv 13 \cdot t_n + 19 \pmod{24} (*).$$

$$t_{n+1} \equiv 13 \cdot t_n + 22 \pmod{24}.$$

$$t_{n+1} \equiv 13 \cdot t_n + 23 \pmod{24} (*).$$

Remark 3: In Theorem 1 in [8], we consider only the symmetric recursive sequences generated by (14). From Theorem 4 and Remark 1, we should add the following three symmetric recursive sequences mod m of the type (14) which are weakly uniformly distributed mod m:

i) $\qquad t_{n+1} \equiv 2 \cdot t_n + 3 \pmod{6}$
$\qquad\qquad \equiv t_n + t_n^{-1} + 3 \pmod{6}.$

ii) $\qquad t_{n+1} \equiv 4 \cdot t_n + 1 \pmod{6}$
$\qquad\qquad \equiv 2 \cdot t_n + 2 \cdot t_n^{-1} + 1 \pmod{6}.$

iii) $\qquad t_{n+1} \equiv 4 \cdot t_n + 5 \pmod{6}$
$\qquad\qquad \equiv 2 \cdot t_n + 2 \cdot t_n^{-1} + 5 \pmod{6}.$

These additions have been made since there has been a miscomprehension in the proof of Theorem 1 in [8].

Remark 4: We have completely enumerated the symmetric recursive sequences mod m that are uniformly distributed in $(\mathbb{Z}/m\mathbb{Z})^*$.

On the other hand, we leave open the possibility that some symmetric recursive sequences mod m other than those in Remark 2 are weakly uniformly distributed mod m, since we have no knowledge concerning non-invertible elements in $\mathbb{Z}/m\mathbb{Z}$. But from the definition of symmetric recursive sequences mod m, Theorem 2 is regarded to be sufficient for our purpose.

Remark 5: The sequence of Fibonacci numbers is, for example, weakly uniformly distributed mod 3. It might be interesting to characterize integers m for which the sequence of Fibonacci numbers is weakly uniformly distributed mod m.

REFERENCES

[1] Bumby, R. T. "A distribution property for linear recurrence of the second order." *Proc. Amer. Math. Soc., 50*, (1975): pp 101-106.

[2] Dickson, L. E. "The analytic representation of substitution on a power of a prime number of letters with a discussion of the linear group." *Ann. Math., 11*, (1896-97): pp 65-120, 161-183.

[3] Kuipers, L., and Shiue, J. S. "On the distribution modulo M of generalized Fibonacci numbers." *Tamkang J. Math., 2*, (1971): pp 181-186.

[4] Kuipers, L., and Shiue, J. S. "A distribution property of the sequence of Fibonacci numbers." *The Fibonacci Quarterly, 10*, (1973): pp 375-376, 392.

[5] Nagasaka, K. "Distribution property of recursive sequences defined by $u_{n+1} \equiv u_n + u_n^{-1}$ (mod m)." *The Fibonacci Quarterly, 22*, (1984): pp 76-81.

[6] Nagasaka, K. "On Benford's law". *Ann. Inst. Stat. Math., 36, Ser. A*, (1984): pp 337-352.

[7] Nagasaka, K. "Weakly uniformly distributed sequences of integers." *Proc. Symp. Res. Inst. Math. Sci. Kyoto Univ., 537*, (1984): pp 1-10.

[8] Nagasaka, K. "Weakly uniform distribution mod m for certain recursive sequences and for monomial sequences." *Tsukuba J. Math., 9*, (1985): pp 159-166.

[9] Narkiewicz, W. "Uniform distribution of sequences of integers." *London Math. Soc. Lec. Note Ser., 56, Cambridge Univ. Press, London*, (1982): pp 202-210.

[10] Narkiewicz, W. "Uniform Distribution of Sequences of Integers in Residue Classes." *Lecture Notes in Math., 1087, Springer-Verlag, Berlin-Heidelberg New York-Tokyo*, (1984).

[11] Niederreiter, H. "Distribution of Fibonacci numbers mod 5^k." *The Fibonacci Quarterly*, *10*, (1972): pp 373-374.

[12] Niven, I. "Uniform distribution of sequences of integers." *Trans. Amer. Math. Soc.*, *98*, (1961): pp 52-61.

Peter Kiss

PRIMITIVE DIVISORS OF LUCAS NUMBERS

INTRODUCTION AND RESULTS

Let $R = \{R_n\}_{n=1}^{\infty}$ be a Lucas sequence defined by fixed rational integers A and B nd by the recursion relation

$$R_n = A \cdot R_{n-1} + B \cdot R_{n-2}$$

or $n > 2$, where the initial values are $R_1 = 1$ and $R_2 = A$. The terms of R are alled Lucas numbers. We shall denote the roots of the characteristic polynomial

$$f(x) = x^2 - Ax - B$$

y α and β. We may assume that $|\alpha| \geq |\beta|$ and the sequence is not degenerate, that s, $AB \neq 0$, $A^2 + 4B \neq 0$ and α/β is not a root of unity. In this case, as it is well-nown, the terms of the sequence R can be expressed as

$$R_n = \frac{\alpha^n - \beta^n}{\alpha - \beta} \quad (n=1, 2, \ldots).$$

If p is a prime and $(p, B) = 1$, then there are terms in the sequence R divisible y p. If $(p, B) = 1$ and $p \mid R_n$ but $p \nmid R_m$ for $m = 1, 2, \ldots, n-1$, then p is called a rimitive prime divisor of R_n. We know that in this case $n \mid (p-1)$ or $n \mid (p+1)$ if p $A^2 + 4B$. Furthermore we say p^e $(e \geq 1)$ is a primitive prime power divisor of R_n ° p is a primitive prime divisor of R_n and $p^e \mid R_n$ but $p^{e+1} \nmid R_n$. We shall denote he product of all primitive prime power divisors of R_n by $\mathbf{R}_n$:

$$\mathbf{R}_n = \prod p^e,$$

here the product is extended over all primitive prime power divisor of R_n. We rite $R = 1$ if R_n has no primitive prime divisor.

29

. N. Philippou et al. (eds.), Applications of Fibonacci Numbers, 29–38.

We prove an asymptotic formula.

Theorem 1: If R is a non-degenerate Lucas sequence, then

$$\sum_{n \leq x} \log | R_n | - \frac{3 \cdot \log |\alpha|}{\pi^2} \cdot x^2 + 0(x \cdot \log x) ,$$

where $\log |\alpha| > 0$ since $|\alpha| \geq |\beta|$ and α/β is not a root of unity, so $|\alpha| > 1$.

We note that J. V. Matiyasevich and R. K. Guy [4] have obtained a similar result for the Fibonacci sequence F ($F_1 = F_2 = 1$, $F_n = F_{n-1} + F_{n-2}$) proving that

$$\pi = \lim_{n \to \infty} \sqrt{(6 \cdot \log M_n)/\log L_n} ,$$

where $M_n = F_1 \cdot F_2 \cdots F_n$ and L_n is the least common multiple of $F_1, F_2, \ldots, F_n$. can be seen that our Theorem 1 also includes this result.

Theorem 1 implies a consequence for the number of primitive prime divisor of the Lucas numbers. Using the prime number formula

$$\pi (x) = \frac{x}{\log x} + 0(x/\log^2 x)$$

and the average order of the Chebyshev's function

$$\emptyset(x) = \sum_{p \leq x} \log p = x + 0(x/\log^2 x)$$

by Theorem 1 we get

Corollary: If $\omega(m)$ denotes the number of distinct prime factors of m, then

$$\sum_{n \leq x} \omega(R_n) < \left[\frac{3 \cdot \log |\alpha|}{2 \cdot \pi^2} + \epsilon \right] \cdot \frac{x^2}{\log x}$$

for any $\epsilon > 0$ and $x > x_0(\epsilon)$.

Some other results can be derived from Theorem 1 and its Corollary. First we recall some known results. We know that there is an absoulte constant n_0 such that R_n has at least one primitive prime divisor for any $n > n_0$ (e.g. see A. Schinzel [7] and C. L. Stewart [9]). We also know results about the greatest prime factors of

ucas numbers (and of general linear recurrences, too); the most of them were
obtained by T. N. Shorey and C. L. Stewart. E.g. in [8] they proved: Let P(n)
denote the greatest prime factor of n and let $q(n) = 2^{\omega(n)}$ ($\omega(n)$ is defined in the
corollary); then for any k with $0 < k < 1/\log 2$ and any integer n (>3) with at most
$\cdot\log \log n$ distinct prime factors, we have

$$P(R_n) > c \cdot (\varphi(n) \cdot \log n)/q(n),$$

where c is a positive number depending only on α, β and k, and φ is the Euler
function. Furthermore, for "almost all" integers n,

$$P(R_n) > \frac{n \cdot \log^2 n}{f(n) \cdot \log \log n} ,$$

where f(n) is any real valued function with $\lim_{n \to \infty} f(n) = \infty$.

We give a result for the greatest primitive prime power factors of Lucas
numbers which will be denoted by $PP(R_n)$.

Theorem 2: Let x and λ ($0 < \lambda < 1$) be real numbers. Let S_x be the set of the
terms R_n of a non-degenerate Lucas sequence for which $n \leq x$ and R_n
has a primitive prime power factor greater than $n^{2-\lambda}$ (i.e. $PP(R_n) > n^{2-\lambda}$).
Then for the cardinality of S_x we have

$$|S_x| > \left[\frac{3 \lambda}{2 \pi^2} - \epsilon\right] x$$

for any $\epsilon > 0$ and $x > x_0$ (ϵ).

Our results give connections among three interesting problems:
1. How large is the greatest prime factor of R_n (or R_n) ?
2. How many primitive prime factors has a Lucas number R_n ?
3. For how many primes does there exist an index n such that $p^e | R_n$
   with $e > 1$?

We have already mentioned some results concerning problem 1.

Problem 2 is also interesting. The existence of a primitive prime factor of terms R_n ($n > n_0$) is known (see K. Zsigmondy [11], A. Schinzel [7] and C. L. Stewart [9]). A. Schinzel [6] proved that there are infinitely many indices n for which R has at least two primitive prime factors (these n's form an arithmetical progression). But we do not know much more about it.

Problem 3 is very hard, we know almost nothing about it. For example, for the Fibonacci sequence F we do not know a prime such that $p^2 | F_n$ would hold if p is a primitive prime divisor of F_n. Or in another special case, $\alpha = 2$ and $\beta = 1$ (A=3 and B=2), the terms of the sequence R have the form $2^n - 1$. The primes p with $p^2 | (2^{p-1} - 1)$ are called Wieferich primes. However up until now we know only two Wieferich primes: 1093 and 3511. As it is well-known, the clearing of the problem of Wieferich primes could lead to the complete solution of Last Fermat's Theorem.

Our results show that if there are only "a few" prime of Wieferich-type ($p^2 | R_{p-1}$ or $p^2 | R_{p+1}$), then "many" Lucas numbers have large prime divisors or have many distinct primitive prime divisors. Theorem 2 implies that the set of indices of Lucas numbers, which have primitive prime power divisors greater than $n^{2-\lambda}$, has positive density. But we do not know whether in general the primes are large or the exponents are. It would be desirable to decide which alternative is true.

Proofs of the Theorems:

We need some auxiliary results for the proofs.

Let $\varnothing_n(\alpha, \beta)$ denote the nth cyclotomic polynomial in α and β for any integer $n > 0$ and any pair α, β of complex numbers, that is

$$\varnothing_n(\alpha, \beta) = \prod_{d|n} (\alpha^{n/d} - \beta^{n/d})^{\mu(d)},$$

where μ is the Moebius function.

From some results of C. L. Stewart (Lemma 6 and 7 in [10]) it follows:

Lemma 1: Let R be a non-degenerate Lucas sequence. Then for n > 12 we have

$$R_n = \lambda_n \cdot \varnothing_n(\alpha, \beta),$$

where $\lambda_n = 1$ or $\lambda_n = 1/P(n/(3,n))$.

We shall use a result on linear forms of logarithms. Let

$$\Lambda = \varsigma_1 \cdot \log \omega_1 + \cdots + \varsigma_r \cdot \log \omega_r ,$$

where the ς's are rational integers and the ω's denote algebraic numbers ($\omega_i \neq 0$ or 1). We assume that not all the ς's are 0 and that the logarithms mean their principal values. Suppose that max $(\varsigma_i) \leq N$ (≥ 4), ω_i has height at most M_i (≥ 4) and that the field generated by the ω's over the raional numbers has degree at most d. Then

Lemma 2: If $\Lambda \neq 0$, then

$$| \Lambda | > N^{-c\Omega \cdot \log \Omega'} ,$$

where

$$\Omega = \log M_1 \cdot \log M_2 \cdots \log M_r$$
$$\Omega' = \Omega/\log M_r$$

and c is an effectively computable constant depending only on r and d ($c=(16rd)^{200r}$). (see A. Baker [2] or A. Baker and C. L. Stewart [3]).

We prove two other lemmas.

Lemma 3: If ω is an algebraic number, $|\omega| = 1$ and ω is not a root of unity, then

$$|1 - \omega^n| > e^{-c \cdot \log n}$$

for any $n \geq 4$, where c is a constant depending only on ω.

Proof: By the conditions of the lemma ω and ω^n are not real complex numbers different from ± 1. Therefore it follows that

$$|1 - \omega^n| > \tfrac{1}{2}|\arg (\omega^n)| = \tfrac{1}{2}|\log \omega^n| = \tfrac{1}{2}|n \cdot \log \omega - t \cdot \log (-1)| ,$$

where the logarithms take their principal values and t is an integer with $|t| < n$. Using Lemma 2, for any $n \geq 4$ we get

$$|1 - \omega^n| > \tfrac{1}{2} \cdot n^{-c'} > e^{-c \cdot \log n}$$

with some c' and c depending only on ω.

Lemma 4: Let α and β be the roots of the characteristic polynomial of a non-degenerate Lucas sequence assuming that $|\alpha| \geq |\beta|$. If $n \geq 4$, then

$$\log |\varnothing_n(\alpha, \beta)| = \varphi(n) \cdot \log |\alpha| + \sum_{d|n} \mu(d) \cdot \log \left| 1 - \left[\tfrac{\beta}{\alpha}\right]^{n/d}\right| ,$$

where φ is the Euler function.

Proof: By the definition of $\varnothing_n(\alpha, \beta)$ we have

$$\log |\varnothing_n(\alpha, \beta)| = \sum_{d|n} \mu(d) \cdot \log |\alpha^{n/d} - \beta^{n/d}| =$$

$$\sum_{d|n} \mu(d) \cdot \tfrac{n}{d} \cdot \log |\alpha| + \sum_{d|n} \mu(d) \cdot \log \left| 1 - \left[\tfrac{\beta}{\alpha}\right]^{n/d}\right| .$$

It implies the lemma since, as it is well-known,

$$\sum_{d|n} \mu(d) \cdot \tfrac{n}{d} = \varphi(n).$$

Now we are already able to prove our theorems. In the proofs we shall use some results of analytic number theory which can be found, for example, in [1] and [5].

Proof of Theorem 1: Since $P(n/(3,n)) \leq n$ by Lemma 1 and 4 we have

$$\sum_{n \leq x} \log |R_n| = \sum_{n \leq x} \log |\varnothing_n (\alpha, \beta)| + 0\left[\sum_{n \leq x} \log n\right] =$$

$$\log |\alpha| \cdot \sum_{n \leq x} \varphi(n) + 0(\log ([x]!)) + E_x ,$$

where

$$E_x = \sum_{n \leq x} \sum_{d|n} \mu(d) \cdot \log \left| 1 - \left[\tfrac{\beta}{\alpha}\right]^{n/d}\right| . \tag{1}$$

It is known that

$$\sum_{n \leq x} \varphi(n) = \frac{3}{\pi^2} x^2 + O(x \cdot \log x)$$

and

$$\log([x]!) = O(x \cdot \log x);$$

therefore we have only to prove that $E_x = O(x \cdot \log x)$. By (1) we can write

$$E_x = \sum_{n \leq x} \sum_{d|n} \mu\left(\frac{n}{d}\right) \cdot \log \left| 1 - \left(\frac{\beta}{\alpha}\right)^d \right| = \tag{2}$$

$$= \sum_{d \leq x} \left[\log \left|1 - \left(\frac{\beta}{\alpha}\right)^d\right| \cdot \sum_{t \leq \frac{x}{d}} \mu(t) \right].$$

If α and β are real numbers, then $|\alpha| > |\beta|$ by the conditions and so $|\beta/\alpha| < 1$. In this case $\log |1 - (\beta/\alpha)^d| = O(1)$ and so $E_x = O(x \cdot \log x)$ directly follows from (2).

If α and β are not real complex numbers, then $|\beta/\alpha| = 1$ and by Lemma 3 we have

$$\log \left| 1 - \left(\frac{\beta}{\alpha}\right)^d \right| = O(\log d). \tag{3}$$

If $\frac{x}{2} < d \leq x$, then $1 \leq \frac{x}{d} < 2$ and $\sum_{t \leq \frac{x}{d}} \mu(t) = 1$. With (2) and (3)

it implies the estimation

$$E_x = E_{x/2} + O\left[\sum_{\frac{x}{2} < d \leq x} \log d \right] = E_{x/2} + O(x \cdot \log x). \tag{4}$$

It is known that

$$\sum_{n \leq y} \mu(n) = O\left[y \cdot e^{-c\sqrt{\log y}} \right];$$

therefore we can write

$$\sum_{n \leq y} \mu(n) = O\left[\frac{y}{\log y} \right].$$

From it, using (3),

$$E_{x/2} = 0 \left[\sum_{d \leq \frac{x}{2}} (\log d) \cdot \frac{x/d}{\log (x/d)} \right] = 0 \left[x \cdot \sum_{d \leq \frac{x}{2}} \frac{\log d}{d} \cdot \frac{1}{\log (x/d)} \right] \tag{5}$$

follows. By Euler's summation formula we get

$$\sum_{d \leq \frac{x}{2}} \frac{1}{\log (x/d)} = 0 \left[\int_1^{x/2} \frac{du}{\log (x/u)} \right] = 0 \left[x \cdot \int_2^x \frac{dv}{v^2 \cdot \log v} \right] = 0(x). \tag{6}$$

Using Abel's identity, (5) and (6) imply the estimation

$$E_{x/2} = 0 \left[x \cdot \frac{\log (x/2)}{x/2} \cdot \sum_{d \leq \frac{x}{2}} \frac{1}{\log (x/d)} \right] = 0(x \cdot \log x). \tag{7}$$

Thus by (4) and (7) we get

$$E_{x/2} = 0(x \cdot \log x)$$

which completes the proof.

Proof of Theorem 2: We may suppose that x is a positive integer. Let ς be a real number satisfying $0 < \varsigma < 3/(2\pi^2)$ and let $R_{n_1}, R_{n_2}, \ldots, R_{n_x}$ be an arrangement of the numbers $\{R_n\}_{n \leq x}$ such that $PP(R_{n_i}) > PP(R_{n_j})$ if i < j. We introduce the notation

$$Q_x = \prod_{x_i > \varsigma x} R_{n_i} .$$

Since an integer n > 2 has at most $n^{c/\log \log n}$ distinct positive divisors (c is a constant) and naturally $\varphi(n) < n$, by Lemmas 1, 3 and 4 we get

$$\log |R_n| < (1 + \epsilon)n \cdot \log |\alpha| \leq (1 + \epsilon)x \cdot \log |\alpha|$$

for any $\epsilon > 0$ and $x \geq n > n_0 (\epsilon)$. From it, using Theorem 1, it follows that

$$\log Q_x > \frac{3 \cdot \log |\alpha|}{\pi^2} x^2 - (1 + \epsilon) \varsigma x^2 \cdot \log |\alpha|$$

and so

$$Q_x > \exp\left\{\left(\frac{3}{\pi^2} - \varsigma - \epsilon\right)x^2 \cdot \log|\alpha|\right\}$$ (8)

for any $\epsilon > 0$ if x is large enough. On the other hand obviously

$$Q_x \leq \left[PP(Q_x)\right]^{\omega(Q_x)},$$ (9)

where $\omega(Q_x)$ denotes the number of distinct prime divisor of Q_x .

The Corollary shows that

$$\omega(Q_x) < \left[\frac{3 \cdot \log|\alpha|}{2\pi^2} + \epsilon\right] \cdot \frac{x^2}{\log x}$$ (10)

so by (8), (9) and (10) we have

$$PP(Q_x) \geq Q_x^{1/\omega(Q_x)} > \exp\left\{2 \cdot \log x \cdot \log|\alpha| \cdot \frac{3 - \pi^2\varsigma - \epsilon'}{3 \cdot \log|\alpha| + \epsilon'}\right\} >$$

$$\exp\left\{\log x \cdot \left(2 - \frac{2\pi^2\varsigma}{3} - \epsilon''\right)\right\} = x^{2 - \frac{2\pi^2\varsigma}{3} - \epsilon''},$$

where $\epsilon' > 0$ and $\epsilon'' > 0$ can be arbitrarily small if x is large. Now let $\lambda = \frac{2\pi^2\varsigma}{3} + \epsilon''$. Then each element of the set

$$S = \left\{R_{n_1}, R_{n_2}, \ldots, R_{n_{[\varsigma x]}}\right\}$$

has a prime power divisor greater than $x^{2-\lambda}$ and

$$|S| = [\varsigma x] > \left[\frac{3\lambda}{2\pi^2} - \epsilon\right]x$$

for any $\epsilon > 0$. The Theorem is proved.

ACKNOWLEDGMENT

The author wishes to thank C. L. Stewart for his valuable suggestions which have improved the original version of these results.

REFERENCES

[1]   Apostol, T.M. "Introduction to Analytic Number Theory." *New York-Heidelberg-Berlin: Springer Verlag,* 1976.

[2]   Baker, A. "The theory of linear forms in logarithms." *Transcedence theory: advances and applications (ed. By A. Baker and D. W. Masser),* London - New York: *Acad. Press,* 1977.

[3]   Baker, A. and Stewart, C. L. "Further aspect of transcedence theory." *Asterisque 41-42* (1977): pp   153-163.

[4]   Matiyasevich, Y. V., and Guy, R. K. "A new formula for π." *Amer. Math. Monthly, 93 No. 8* (1986): pp 631-635.

[5]   Prachar, K. "Primzahlverteilung." *Berlin-Göttingen-Heidelberg: Springer Verlag,* 1957.

[6]   Schinzel, A. "On primitive prime factors of Lehmer numbers I." *Acta Arithm. 8* (1963): pp   213-223.

[7]   Schinzel, A. "Primitive divisors of the expression $A^n - B^n$ in algebraic number fields." *J. reine angew. Math. 268/269* (1974): pp   27-33.

[8]   Shorey, T. N. and Stewart, C. L. "On divisors of Fermat, Fibonacci, Lucas and Lehmer numbers II." *J. London Math. Soc. (2) 23* (1981): pp   17-23.

[9]   Stewart, C. L. "Primitive divisors of Lucas and Lehmer numbers." *Transcendence theory: advances and applications (ed. by A. Baker and D. W. Masser), London - New York: Acad. Press,* 1977.

[10] Stewart, C. L. "On divisors of Fermat, Fibonacci, Lucas and Lehmer numbers." *Proc. London Math. Soc. 35* (1977): pp   425-447.

[11] Zsigmondy, K. "Zur Theorie der Potenzreste." *Monatsh. Math. 3* (1982): pp 265-284.

H. T. Freitag and G. M. Phillips

A CONGRUENCE RELATION FOR A LINEAR RECURSIVE SEQUENCE OF ARBITRARY ORDER

We consider the sequence $\{T_n\}_0^\infty$ defined by

$$T_{n+m+1} = \sum_{r=0}^{m} a_r\, T_{n+r}, \quad n \geq 0, \tag{1}$$

with initial conditions

$$T_r = c_r, \quad 0 \leq r \leq m,$$

where the a_r and c_r are integers and $a_0 \neq 0$. (If the c_r are all zero, $\{T_n\}_0^\infty$ becomes the null sequence. In this case Theorems 1 and 2 below are trivial.) In (1) $m \geq 0$ is a fixed integer. We referee to (1) as an (m+1)th order recurrence relation or an (m+1)th order difference equation. Thus $\{T_n\}$ is an integer sequence. The purpose of our present paper is to generalize results which we obtained [2] for a sequence $T_n\}$ defined by a second order recurrence relation (m = 1 in (1)), the Fibonacci and Lucas sequences being important special cases. (The case m = 0 is trivial.)

We will assume that the characteristic polynomial associated with (1),

$$x^{m+1} - \sum_{r=0}^{m} a_r\, x^r, \tag{2}$$

has distinct zeros $\alpha_0, \alpha_1, \ldots, \alpha_m$, and we note that, as a consequence of the restriction $a_0 \neq 0$, each $\alpha_j \neq 0$. Then we may write

$$T_n = \sum_{j=0}^{m} A_j\, \alpha_j^n, \quad n \geq 0, \tag{3}$$

where the A_j are uniquely determined by the values of c_r and a_r. In general, the A_j and α_j will be complex.

Let p denote any prime. We define

A. N. Philippou et al. (eds.), Applications of Fibonacci Numbers, 39–44.
© 1988 by Kluwer Academic Publishers.

$$E(n,p) = T_{n+(m+1)p} - \sum_{r=0}^{m} a_r \, T_{n+rp} , \qquad n \geq 0.$$

Note that, on extending the definition of $E(n,p)$ to $p = 1$, we have $E(n,1) = 0$. We will prove that

$$E(n,p) \equiv 0 \qquad (\bmod \ p)$$

for any choice of prime p.

Although E depends also on the a_r and the c_r, we are concerned only with the dependence of E on n and p, for a given sequence $\{T_n\}$ defined as above.

On substituting for each T_n from (3), we obtain

$$E(n,p) = \sum_{j=0}^{m} A_j \, \alpha_j^{n} \left[\alpha_j^{(m+1)p} - \sum_{r=0}^{m} a_r \, \alpha_j^{rp} \right] . \tag{4}$$

Since each α_j is a zero of the polynomial (2), we may write

$$\alpha_j^{(m+1)p} = \left[\sum_{r=0}^{m} a_r \, \alpha_j^{r} \right]^{p} . \tag{5}$$

We now apply the multinomial theorem to the expression on the right of (5). The number of terms in this expansion depends on the number of non-zero coefficients a_r. For notational simplicity, we will assume that each $a_r \neq 0$. This does not affect the generality of our results. Thus we obtain from (5)

$$\alpha_j^{(m+1)p} = \sum \frac{p!}{n_0! \, \ldots \, . \, n_m!} \prod_{r=0}^{m} (a_r \, \alpha_j^{r})^{n_r}, \tag{6}$$

where the summation is taken over all non-negative integers $n_0, \ldots, n_m$ such that $n_0 + \ldots + n_m = p$.

We first prove the following result.

Lemma: If p is any prime and $n_0, \ldots, n_m$ are non-negative integers such that $n_0 + \ldots + n_m = p$, then the multinomial coefficient satisfies

$$\frac{p!}{n_0! \ldots n_m!} \equiv 0 \pmod{p}, \tag{7}$$

unless any $n_r = p$, when the coefficient has the value 1.

Proof: The case $m = 0$ is trivial, as is the case when $m > 0$ and any $n_r = p$. It suffices to consider the case where $m > 0$ and $n_r < p$ for all r. Then $p \mid p!$ and $p \nmid n_r!$ and the lemma follows.

We now split the summation on the right of (6) into two parts, writing

$$\alpha_j^{(m+1)p} = \Sigma^* \frac{p!}{n_0! \ldots n_m!} \prod_{r=0}^{m} (a_r \, \alpha_j^r)^{n_r}$$

$$+ \sum_{r=0}^{m} a_r^p \, \alpha_j^{rp}, \tag{9}$$

where Σ^* denotes the summation over all n_r such that $n_0 + \ldots + n_m = p$ and $0 \le n_r < p$, $0 \le r \le m$. (Note that the n_r are not ordered in any way.) Thus Σ^* simply omits from the summation in (6) those choices of $n_0, \ldots, n_m$ for which any $n_r = p$.

The lemma shows that each multinomial coefficient in Σ^* satisfies (7). On substituting (9) into (4), we obtain

$$E(n,p) = \sum_{j=0}^{m} A_j \, \alpha_j^n \, \Sigma^* \frac{p!}{n_0! \ldots n_m!} \prod_{r=0}^{m} (a_r \, \alpha_j^r)^{n_r}$$

$$+ \sum_{j=0}^{m} A_j \, \alpha_j^n \sum_{r=0}^{m} (a_r^p - a_r) \cdot \alpha_j^{rp}$$

$$= F(n,p) + G(n,p),$$

say. Rearranging the order of summations in the first double sum, and using (3), we obtain

$$F(n,p) = \Sigma^* \left[\frac{p!}{n_0! \ldots n_m!} \left[\prod_{r=0}^{m} a_r^{n_r} \right] T_{N^*} \right],$$

where $N^* = n + \sum_{r=0}^{m} r\, n_r$. Since each term in Σ^* is divisible by p, it follows that

$$F(n,p) \equiv 0 \quad (\text{mod } p).$$

Again, reversing the order of the summations, we have

$$G(n,p) = \sum_{r=0}^{m} (a_r^p - a_r) T_{n+rp}$$

and, by Fermat's theorem,

$$G(n,p) \equiv 0 \quad (\text{mod } p).$$

This establishes our congruence result for $E(n,p)$ which we now state formally.

Theorem 1: Let the sequence $\{T_n\}$ be defined by the (m+1)th order recurrence relation

$$T_{n+m+1} = \sum_{r=0}^{m} a_r\, T_{n+r}, \qquad a_0 \neq 0, \tag{10}$$

with arbitrary starting values $T_r = c_r$, $0 \le r \le m$, where the a_r and c_r are integers. Further, let the characteristic polynomial of the above recurrence relation have distinct zeros. Then, for any choice of prime p,

$$T_{n+(m+1)p} - \sum_{r=0}^{m} a_r\, T_{n+rp} \equiv 0 \quad (\text{mod } p), \tag{11}$$

for $n = 0,1,\ldots$.

This generalizes an earlier result of the present authors [2], obtained for the special case of (10) with $m = 1$, that is, for a sequence defined by a general second order recurrence relation. The authors [2] also obtained a stronger result for this case ($m = 1$) with the restriction $p \ge 5$. For convenience, we restate this here.

Theorem 2: Let the sequence $\{T_n\}$ be defined by

$$T_{n+2} = a_0 T_n + a_1 T_{n+1}, \qquad a_0 \neq 0, \tag{12}$$

where $T_0 = c_0$, $T_1 = c_1$ and the a_r and c_r are arbitrary integers. Then, if $p \geq 5$ is a prime,

$$T_{n+2p} - a_0 T_{n+p} - a_1 T_n \equiv 0 \quad (\text{mod } 2p),$$

for $n = 0, 1, \ldots$.

Theorem 2, in turn, generalizes a still earlier result, due to the first author [1]:

$$F_{n+10} \equiv F_{n+5} + F_n \quad (\text{mod } 10),$$

for all $n \geq 0$.

It should also be pointed out that in Theorem 2, unlike Theorem 1, we do not ask that the roots of the characteristic polynomial be distinct.

In view of Theorem 2, it is natural to ask whether this stronger result, with mod 2p in place of mod p, applies to the sequence $\{T_n\}$ defined by a general (m+1)th order recurrence relation, at least for p sufficiently large (possibly depending on m). In the proof of Theorem 1 above, we can easily see that

$$G(n,p) \equiv 0 \quad (\text{mod } 2p),$$

at least for $p \geq 3$, since

$$a_r^p - a_r \equiv 0 \quad (\text{mod } 2p)$$

for any a_r. However, we are unable to discover whether such a stronger result holds for $F(n,p)$.

REFERENCES

[1] Freitag, H. T. "A Property of Unit Digits of Fibonacci Numbers and a 'Non-Elegant' Way of Expressing G^n." *Proceedings of the First International Conference on Fibonacci Numbers and Their Applications, University of Patras, Greece, August 27-31,* 1984.

[2] Freitag, H. T. and Phillips, G. M. "A Congruence Relation for Certain Recursive Sequences." *Fibonacci Quarterly 24,* (1986): pp   332-335.

Colin M. Campbell, Edmund F. Robertson and Richard M. Thomas

FIBONACCI NUMBERS AND GROUPS

1. GROUP PRESENTATIONS

The problems discussed in this paper arise from the theory of group resentations. In this section, we give a brief review of this subject, or, at least, those aspects of it which are relevant to the present paper. In the remaining ections we discuss links, occurring in our work over a number of years, between this topic and the Fibonacci and Lucas sequences of numbers (f_n) and (g_n).

Group presentations were originally introduced at the end of the last century, particularly in relation to problems in topology. Informally, a presentation or a group G consists of a set $\{g_i \mid i \in I\}$ of elements which generate G, i.e. every lement of G can be expressed as a product of the g_i and g_i^{-1}, together with a set $r_j \mid j \in J)$ of relators which determine the structure of G, i.e. the r_j are words in he g_i and g_i^{-1}, and the multiplication in G may be determined solely by the usual roup axioms and the relations $r_j = 1$ $(j \in J)$. We shall only be interested in finite resentations, i.e. in cases where both I and J are finite. A group with a finite resentation may be either finite or infinite. A presentation is then usually written n the form:

$$< g_1, g_2, \ldots, g_n \mid r_1 = 1, r_2 = 1, \ldots, r_m = 1 >,$$

r even:

$$< g_1, g_2, \ldots, g_n \mid s_1 = t_1, s_2 = t_2, \ldots, s_m = t_m >,$$

vhere the relation $s_i = t_i$ is equivalent to the relation $s_i t_i^{-1} = 1$. For example, both:

$$< a \mid a^{11} = 1 >$$

and:

$$< a, b, c, d, e \mid ab = c, bc = d, cd = e, de = a, ea = b >$$

are presentations for the cyclic group C_{11} of order 11 (though, in the second case,

45

A. N. Philippou et al. (eds.), Applications of Fibonacci Numbers, 45–60.
© 1988 by Kluwer Academic Publishers.

this is not immediately obvious). On the other hand, one may show that the group with presentation:

$$< a, b, c, d, e, f \mid ab = c, bc = d, cd = e, de = f, ef = a, fa = b >$$

is infinite. We shall mention these examples again in section 2.

More formally, the group G with presentation:

$$< g_1, g_2, \ldots , g_n \mid r_1 = 1, r_2 = 1, \ldots , r_m = 1 >$$

is defined to be the factor group F/N, where F is the free group on $\{g_1, g_2, \ldots g_n\}$, and N is the smallest normal subgroup of F containing the elements $\{r_1, r_2, \ldots r_m\}$. For a more detailed account of group presentations, the reader is referred to (20, 21, 25, 26, 29) in particular.

One term we shall use in the paper is the deficiency of a presentation. Given a presentation:

$$< g_1, g_2, \ldots , g_n \mid r_1 = 1, r_2 = 1, \ldots, r_m = 1 >,$$

the deficiency of the presentation is defined to be n - m. The deficiency of a group is defined to be the maximum of the deficiencies of its presentations. It is well known (see, for example, 20, 21) that a finite group has no presentation with more generators than relations, so that any finite group has non-positive deficiency. The groups we shall be interested in will have deficiency 0 or -1. (See 22 for a general survey of groups of deficiency 0.)

Lastly, we should note that, for many problems involving group presentations there is no general and effective method for resolving the question. To put this more formally, let S be the set of all finite presentations, and let T(P) be the subset of S of all presentations such that the group defined by that presentation has property P. Suppose that S is mapped, via a Gödel numbering, onto the integers Z. Then, for some properties P, T(P) is not mapped onto a subset of Z with an effectively computable characteristic function. Among such properties P are the properties of finiteness and triviality. So there is no general and effective method

' determining whether the group defined by a presentation is finite, or even whether it is trivial. For a fuller account of these topics, and proofs of these two results, the reader is referred to (28).

2. FIBONACCI GROUPS

In 1965, Conway posed the problem of showing that the group with presentation:

$$< a, b, c, d, e \mid ab = c, bc = d, cd = e, de = a, ea = b >$$

cyclic of order 11 (15). This was soon settled, and the problem extended to considering the groups G_n with presentations:

$$< a_1, a_2, \ldots, a_n \mid a_1a_2 = a_3, a_2a_3 = a_4, \ldots, a_{n-1}a_n = a_1, a_na_1 = a_2 >$$

6), Conway's group then being G_5. In (16), it was shown that G_1 and G_2 are trivial, G_3 is the quaternion group Q_8 of order 8, G_4 the cyclic group of order 5, G_5 indeed the cyclic group of order 11, and G_6 (also mentioned in section 1) infinite. Since then, G_7 has been shown to be cyclic of order 29 (1, 14, 17), G_8 and G_{10} to be finite (1) , and G_n to be infinite for $n \geq 11$ (24). Only the nature of G_9 remains to be determined, though its order, if finite, is at least 152.5^{741} (18).

The groups G_n came to be known as Fibonacci groups. But why the mention of Fibonacci? There are two reasons. Firstly, G_n has the Fibonacci-type relations $a_ia_{i+1} = a_{i+2}$, and, secondly, if we consider the group $A_n = G_n/G_n'$ (where G_n' is the derived group of G_n, i.e. the subgroup generated by all the commutators [x, y] = $x^{-1}y^{-1}xy$), then $\mid A_n \mid = g_n - 1 - (-1)^n$ (16).

The groups G_n are examples of cyclically presented groups. To explain this term, let F_n be the free group of rank n with generators $a_1, a_2, \ldots, a_n$, and let Θ be the automorphism of F_n defined by:

$$a_1^{\Theta} = a_2, a_2^{\Theta} = a_3, \ldots, a_{n-1}^{\Theta} = a_n, a_n^{\Theta} = a_1.$$

If w is a word in F_n, let N be the smallest normal subgroup of F_n containing w, w^Θ $w^{\Theta^2}, \ldots, w^{\Theta^{n-1}}$, and then let $G_n(w)$ denote the group F_n/N. Then $G_n(w)$ h presentation:

$$< a_1, a_2, \ldots, a_n \mid w = w^\Theta = w^{\Theta^2} = \ldots = w^{\Theta^{n-1}} = 1 >,$$

and is said to be cyclically presented. The groups G_n mentioned above are t groups $G_n(a_1 a_2 a_3^{-1})$ in this notation. More generally, the Fibonacci group $F(r, n)$ defined to be $G_n(a_1 a_2 \ldots a_r a_{r+1}^{-1})$, where the subscripts are reduced modulo n, t groups G_n then being $F(2, n)$. For a general survey of the Fibonacci groups, see ([1] [20], [21], [23]). In the same way as the Fibonacci sequence itself has been generalize so various generalizations of the Fibonacci groups have been considered, includi the groups:

$$F(r, n, k) = G_n(a_1 a_2 \ldots a_r a_{r+k}^{-1}) \quad (r \geq 2, k \geq 0),$$

(see [4], [5], [6], [13], [19] for example), and the groups:

$$H(r, n, s) = G_n((a_1 a_2 \ldots a_r) (a_{r+1} a_{r+2} \ldots a_{r+s})^{-1}) \quad (r > s \geq 1),$$

(see [2], [8], [13] for example). Clearly, the groups $F(r, n, 1)$ and $H(r, n, 1)$ are eac isomorphic to $F(r, n)$.

Whilst on the subject of cyclically presented groups, we note that cyclically presented finite group must have deficiency zero, since a presentation the form:

$$< a_1, a_2, \ldots, a_n \mid w = w^\Theta = w^{\Theta^2} = \ldots = w^{\Theta^{n-1}} = 1 >$$

has n generators and n relations, and no presentation for a finite group can hav more generators than relations.

3. THE GROUPS X(n) AND Y(n)

Of the few groups of Fibonacci type known to be finite and non-metacyclic, here are, surprisingly, two non-isomorphic groups of order 1512. These are $F(3, 6)$ and the group with presentation:

$$< a, b, c, d, e, f \mid ace = d, \; bdf = e, \; cea = f, \; dfb = a, \; eac = b, \; fbd = c >,$$

e. the groups $G_6(a_1a_2a_3a_4^{-1})$ and $G_6(a_1a_3a_5a_4^{-1})$ in the notation of the previous section $5, 7$). The second group has the two-generator two-relation presentation:

$$< a, b \mid (abab^{-1})^2 = ba^{-1}bab^{-1}a, \; ab^2a^{-1}ba^2b^{-1} = 1 >.$$

To gain insight into this group, the groups $Y(n)$ with presentations:

$$< a, b \mid (abab^{-1})^n = ba^{-1}bab^{-1}a, \; ab^2a^{-1}ba^2b^{-1} = 1 >$$

$n \in Z$) were considered in ($\underline{10}$). It was possible to show, using a computer, that, for small n, n coprime to 3, the relation $ab^2aba^2b = 1$ holds, and the groups $X(n)$ with presentations:

$$< a, b \mid (abab^{-1})^n = ba^{-1}bab^{-1}a, \; ab^2aba^2b = 1 >$$

$n \in Z$) were also considered. If we let $T(m)$ denote the group with presentation:

$$< x, t \mid xt^2xtx^2t = 1, \; xt^{m+1} = t^2x^2 >,$$

then the following results were obtained:

<u>Theorem (3.1)</u>: $X(n)$ has a subgroup of index $\mid 2n + 3 \mid$ isomorphic to $T(2n)$.

<u>Theorem (3.2)</u>: The group $T(m)$ has order the modulus of:

    (i) $8ng_n^3$, m = 2n, n even.

    (ii) $40nf_n^3$, m = 2n, n odd.

    (iii) $4mg_m$, m odd, (m, 3) = 1.

    (iv) $8mg_m$, m odd, (m, 3) = 3.

Combining Theorems (3.1) and (3.2) yields:

<u>Theorem (3.3)</u>: The group X(n) has order the modulus of:

    (i) $8n(2n + 3)g_n^3$ if n is even.

    (ii) $40n(2n + 3)f_n^3$ if n is odd.

We shall return to the groups T(m) in the next section. The groups Y(n) wer shown to be infinite if (n, 3) = 3 (<u>10</u>). It was conjectured (<u>10</u>, <u>12</u>) that Y(n) isomorphic to the group $\overline{X}(n)$ with presentation:

$$< a, b \mid (abab^{-1})^n = ba^{-1}bab^{-1}a, ab^2aba^2b = 1, ab^2a^{-1}ba^2b^{-1} = 1 >$$

if (n, 3) = 1. $\overline{X}(n)$ is a factor group of X(n) by a central subgroup of order 2. Th results of (<u>10</u>, <u>12</u>) show that the conjecture is true for $-10 \leq n \leq 10$ with (n, 3) 1.

4. THE GROUPS $F^{a,b,c}$

Consider now the group $F^{a,b,c}$ with presentation:

$$< x, y \mid x^2 = 1, xy^axy^bxy^c = 1 >.$$

These groups were of interest because of their relevance to the search fo trivalent zero-symmetric graphs. The structure of these groups depends heavily o that of the groups $H^{a,b,c}$ with presentation:

$$< x, y \mid x^2 = 1, y^{2n} = 1, xy^a xy^b xy^c = 1 >,$$

where $n = a + b + c$. These groups are discussed in (3, 9, 11). One result of present interest is the following (3, 11):

Theorem (4.1): Suppose $(a, b) = 1$.

    (i)   If $(a-b, 3) = 3$, then $\mid F^{a,b,-2a} \mid = 2 \mid H^{a,b,-2a} \mid = 4n(g_n + 1 + (-1)^n)$.

    (ii) If $(a-b, 3) = 1$, then $\mid F^{a,b,-2a} \mid = \mid H^{a,b,-2a} \mid = 2n(g_n + 1 + (-1)^n)$.

The groups $F^{a,b,c}$ are related to the groups $T(m)$ with presentations:

$$< x, t \mid xt^2 xtx^2 t = 1, xt^{m+1} = t^2 x^2 >$$

discussed in section 3. If we put $a = xt$, $b = t$, then $T(m)$ has presentation:

$$< a, b \mid b^{-1} a^2 b = a^{-2}, abab^{-2} ab^{m+1} = 1 >.$$

Adding the relation $a^2 = 1$, i.e. factoring out the normal subgroup $<a^2>$, yields:

$$< a, b \mid a^2 = 1, abab^{-2} ab^{m+1} = 1 >,$$

which is a presentation for $F^{1,-2,m+1}$.

    In an attempt to shed light on the involvement of the Fibonacci and Lucas numbers in the classes $F^{a,b,c}$ and $H^{a,b,c}$ of groups, the related class of groups $N^{a,b,c}$ with presentation:

$$< x, y \mid y^{-1} x^2 y = x^{-2}, y^{2n} = 1, xy^a xy^b xy^c = 1 >,$$

where $n = a + b + c$, was considered in (11). The subgroup $<x^2>$ of $N^{a,b,c}$ is normal, and $N^{a,b,c}/ <x^2>$ is isomorphic to $H^{a,b,c}$. Since the order of $H^{a,b,c}$ had been determined in (3), the problem was to determine the order of $<x^2>$. It is easy to show that $N^{a,b,c}$ is isomorphic to $N^{-c,-b,-a}$, so that we may assume that $n \geq 0$. The case n = 0 yields an infinite group, so assume n $\geq$ 1. Fibonacci numbers occur when we consider the case n even, a and b odd.

Let $L^{a,b,c}$ be the derived group of $N^{a,b,c}$, so that $N^{a,b,c}/L^{a,b,c}$ is isomorphic to C_{2n}. One may derive a presentation for $L^{a,b,c}$ of the form:

$$< Y_1, Y_2, \ldots, Y_n, T \mid Y_{i+a+b} Y_i = Y_{i+a} T^{\delta_i}, [T, Y_i] = 1, Y_i = Y_{i+n}^{-1} \ (1 \leq i \leq 2n) >$$

where the δ_i are determined in terms of a, b, c, and the subscripts are reduced mod 2n. For any integers s and t, we have the commutator relations:

$$\left[Y_i, Y_{i+sa+tb} \right] = T^{\mu(i,s,t)},$$

where $\mu(i, s, t) = (-1)^{i+s} f_{s+t}$. From this, one deduces:

Theorem (4.2): Suppose that n is a positive even integer and that a and b are odd. Let $\alpha = (a-b)/2$, $\beta = n/(2(n, a))$, $\gamma = n/(2(n, b))$. Then x^2 has order:

    (i) $(f_\alpha, f_\beta, f_\gamma) = f_{(\alpha, \beta, \gamma)}$ if α and $n/2$ are odd.

    (ii) $(g_\alpha, f_\beta, f_\gamma) = (g_\alpha, f_{(\beta, \gamma)})$ if α is even, $n/2$ odd.

    (iii) $(f_\alpha, g_\beta, g_\gamma)$ if α is odd, $n/2$ even.

    (iv) $(g_\alpha, g_\beta, g_\gamma)$ if α and $n/2$ are even.

From this, we see that we may choose a, b, c so that x^2 has order f_{2m+1} or g_{2m} for any m. If we take c = -2b, so that n = a - b, then, provided (a, b) = 1, x^2 has order $f_{(a-b)/2}$ if (a - b)/2 is odd, and order $g_{(a-b)/2}$ if (a - b)/2 is even.

5. RECENT RESULTS

Let us now return to the Fibonacci groups F(r, n) and the generalized Fibonacci groups H(r, n, s) introduced in section 2. The automorphism $\ominus$ of a free group F_n with generators $a_1, a_2, \ldots, a_n$ induces an automorphism on a cyclically presented group $G_n(w)$, so that we may form a semi-direct product E(r, n) of F(r, n) with a cyclic group <t> of order n. If we let x denote ta_1^{-1}, then E(r, n) has resentation:

$$< x, t \mid xt^r = tx^r, t^n = 1 >.$$

$F(r, n)$ then has a homomorphic image with presentation:

$$< x, t \mid xt^r = tx^r, t^n = x^n = 1 >,$$

which admits an automorphism α of order 2 interchanging x and t. Adjoining α, we get the group $G(r, n)$ with presentation:

$$< x, \alpha \mid x^n = \alpha^2 = \alpha x \alpha x^r \alpha x^{-r} \alpha x^{-1} = 1 >.$$

In (30), it was shown that, if $d = (r + 1, n)$, then $G(r, n)$ is infinite whenever $d > 3$, or $d = 3$ with n even, and so we have:

Theorem (5.1): $F(r, n)$ is infinite if either:

    (i) $(r + 1, n) > 3$, or:

    (ii) $(r + 1, n) = 3$ with n even.

This work was extended in (13), where we have:

Theorem (5.2): $H(r, n, s)$ is infinite if one of the following conditions holds:

    (i) $(r + s, n) > 3$,

    (ii) $(r + s, n) = 3$ with $r + s > 3$,

    (iii) $(r + s, n) > 1$ with $(r, s) > 1$.

Putting $s = 1$ in Theorem (5.2) (ii) extends Theorem (5.1) to cover the case $(r + 1, n) = 3$ with $r > 2$. Since the group $F(r, n)$ is infinite whenever $G(r, n)$ is infinite, the question arises as to which of the groups $G(r, n)$ are finite and which infinite. We shall show that, even with the groups $G(r, n)$, the Lucas numbers g_n make an appearance. More precisely, we shall prove:

<u>Theorem (5.3)</u>: (i) G(2, 2) is elementary abelian of order 4.

(ii) G(2, 4) is metabelian of order 40.

(iii) G(2, n) is infinite for n even, n $\geq$ 6.

(iv) G(2, n) is metabelian of order $2ng_n$ for n odd.

<u>Proof</u>: Since our primary interest here is in the Fibonacci and Lucas numbers, we make only a brief reference to the cases where n is even. If n = 2, we have the group with presentation:

$$< a, b \mid a^2 = b^2 = abab^{-1} = 1 >,$$

which is easily seen to be a presentation for $C_2 \times C_2$. If n = 4, we have:

$$< a, b \mid a^2 = b^4 = abab^2ab^{-2}ab^{-1} = 1 >.$$

Let N = < a, bab^{-1}, b^2ab^{-2}, b^3ab^{-3} >. Then N is seen to be a normal subgroup of order 10, and so G to have order 40. If n > 4, we let c = aba, and then M be the normal subgroup < b, c>. We may show that M has presentation:

$$< b, c \mid b^n = 1, cb^2 = bc^2 >,$$

so that M is isomorphic to E(2, n). Since F(2, n) is infinite for n even, n $\geq$ 6, by the results of section 2, we have that G(2, n) is infinite in this case.

Turning to the case of n odd, we describe a method of proof similar to that used to prove many of the theorems quoted earlier. G(2, n) has presentation:

$$< a, b \mid a^2 = b^n = abab^2ab^{-2}ab^{-1} = 1 >.$$

Let x = $ab^{-1}ab$, y = $abab^{-1}$, n = 2k - 1. Then the relation $abab^2ab^{-2}ab^{-1} = 1$ gives:

$$ba.b^2.ab^{-1} = (ab)^2$$

and so,

$$(ab)^{2k} = ba.b^{2k}.ab^{-1} = babab^{-1},$$

that is:

$$(ab)^n = (ab)^{2k}(ab)^{-1} = babab^{-2}a.$$

Now $(ab)^{2n} = ba.b^{2n}.ab^{-1} = 1$, so that:

$$(babab^{-2}a)^2 = 1,$$

that is:

$$(bab^{-1}a.ab^{-1}ab.a)^2 = 1,$$

which is:

$$(y^{-1}xa)^2 = 1.$$

Now $axa = b^{-1}aba = x^{-1}$, $aya = bab^{-1}a = y^{-1}$, so that $(y^{-1}xa)^2 = 1$ implies:

$$y^{-1}xyx^{-1} = 1.$$

Also note that $bxb^{-1} = y^{-1}$ and

$$byb^{-1} = bab^{-1}a.ab^2ab^{-2}$$
$$= bab^{-1}a.b^{-1}aba$$
$$= y^{-1}x^{-1}.$$

We have now shown that $H = <x, y>$ is a normal abelian subgroup of $G(2, n)$, and hence that $G(2, n)$ is metabelian. Since $|G(2, n)/H| = 2n$, it remains to show that $|H| = g_n$.

We obtain a presentation for H from the modified Todd-Coxeter coset enumeration algorithm (see 27 for example). Define cosets and obtain coset representatives as follows:

1 a = 2.	Deduce 2 a = 1.
1 b^{-1} = 3.	
3 a = 4.	Deduce 4 a = 3, 4 b = x^{-1} 2.
1 b = 5.	

$$5 \ a = 6. \qquad \text{Deduce } 6 \ a = 5, \ 2 \ b = y \ 6.$$

In general, we define:

$$(2i - 1) \ a = 2i \qquad (1 \leq i \leq n),$$

and deduce:

$$(2i) \ a = 2i - 1 \qquad (1 \leq i \leq n)$$

from the relation $a^2 = 1$. Also, we define:

$$(2i - 1) \ b = 2i + 1 \qquad (3 \leq i \leq n - 1).$$

From the relation $b^n = 1$, we get:

$$(2n - 1) \ b = 3.$$

In the rest of this proof we use the fact that x and y commute.

Now, from the relation $ab^{-1}abab^2ab^{-2} = 1$, we get, from 1 $ab^{-1}abab^2ab^{-2} = 1$:

$$6 \ b = x^{-1}y^{-1} \ 8.$$

Also using the relations:

$$(2i - 1) \ ab^{-1}abab^2ab^{-2} = 2i - 1 \qquad (3 \leq i \leq n - 2)$$

successively gives:

$$(2i + 2) \ b = x^{(-1)^{i+1}f_{i-1}} \ y^{(-1)^{i+1}f_i} \ (2i + 4) \qquad (3 \leq i \leq n - 2).$$

The relation $2b^n = 2$ gives $(2n) \ b = x^\alpha y^\beta \ 4$ where:

$$f_0 - f_1 + f_2 - f_3 + \ldots \ldots + f_{n-3} + \alpha - 1 = 0,$$
$$f_1 - f_2 + f_3 - f_4 + \ldots \ldots + f_{n-2} + \beta = 0.$$

It follows by induction that:

$$\alpha = -f_{n-4} + 2,$$
$$\beta = -f_{n-3} - 1.$$

A presentation for the subgroup is obtained from:

$$3\ ab^{-1}abab^2ab^{-2} = 3,$$
$$(2n - 3)\ ab^{-1}abab^2ab^{-2} = 2n - 3,$$
$$(2n - 1)\ ab^{-1}abab^2ab^{-2} = 2n - 1,$$

since it is straightforward to check that:

$$(2i)\ ab^{-1}abab^2ab^{-2} = 2i \quad (1 \leq i \leq n)$$

give no further relations for the subgroup.

From $3\ ab^{-1}abab^2ab^{-2} = 3$:

$$x^{f_{n-4} - 2}\ y^{f_{n-3} + 1}\ x^{-1}\ y = 1.$$

From $(2n - 3)\ ab^{-1}abab^2ab^{-2} = 2n - 3$:

$$x^{f_{n-4}}\ y^{f_{n-3}}\ x^{f_{n-3}}\ y^{f_{n-2}}\ x^{-f_{n-4} + 2}\ y^{-f_{n-3} - 1} = 1.$$

From $(2n - 1)\ ab^{-1}abab^2ab^{-2} = 2n - 1$:

$$x^{-f_{n-3}}\ y^{-f_{n-2}}\ x^{-f_{n-4} + 2}\ y^{-f_{n-3} - 1}\ x^{-1} = 1.$$

So the presentation for H is:

$$< x, y \mid x^{f_{n-4} - 3}\ y^{f_{n-3} + 2} = x^{f_{n-3} + 2}\ y^{f_{n-2} - 1} = x^{-f_{n-2} + 1}\ y^{-f_{n-1} - 1} = 1 >.$$

The first relation is redundant via the other two. Thus H has order:

$$\begin{vmatrix} f_{n-3} + 2 & f_{n-2} - 1 \\ \\ f_{n-2} - 1 & f_{n-1} + 1 \end{vmatrix}$$

$$= (f_{n-3}\, f_{n-1} - f_{n-2}^{\,2} + 1) + f_{n-3} + 2\, f_{n-1} + 2\, f_{n-2}$$
$$= 0 + (f_{n-3} + f_{n-2}) + (f_{n-2} + f_{n-1}) + f_{n-1}$$
$$= f_{n-1} + f_n + f_{n-1}$$
$$= g_n$$

as required.

REFERENCES

[1]   Brunner, A. M. "The determination of Fibonacci groups." *Bull. Austral. Math. Soc. 11* (1974): pp 11-14.

[2]   Brunner, A. M. "On groups of Fibonacci type." *Proc. Edinburgh Math. Soc.* 20 (1977): pp 211-213.

[3]   Campbell, C. M., Coxeter, H. S. M. and Robertson, E. F. "Some families of finite groups having two generators and two relations." *Proc. Roy. Soc. London* 357A (1977): pp 423-438.

[4]   Campbell, C. M. and Robertson, E. F. "The orders of certain metacyclic groups." *Bull. London Math. Soc. 6* (1974): pp 312-314.

[5]   Campbell, C. M. and Robertson, E. F. "Applications of the Todd-Coxeter algorithm to generalized Fibonacci groups." *Proc. Roy. Soc. Edinburgh* 73A (1974/5): pp 163-166.

[6]   Campbell, C. M. and Robertson, E. F. "On metacyclic Fibonacci groups." *Proc. Edinburgh Math. Soc.* 19 (1975): pp 253-256.

[7]   Campbell, C. M. and Robertson, E. F. "A note on Fibonacci type groups." *Canad. Math. Bull.* 18 (1975): pp 173-175.

[8]   Campbell, C. M. and Robertson, E. F. "On a class of finitely presented groups of Fibonacci type." *J. London Math. Soc.* 11 (1975): pp 249-255.

9]   Campbell, C. M. and Robertson, E. F. "Classes of groups related to $F^{a,b,c}$."
     Proc. Roy. Soc. Edinburgh 78A (1978): pp 209-218.

10] Campbell, C. M. and Robertson, E. F. "Deficiency zero groups involving
     Fibonacci and Lucas numbers." Proc. Roy. Soc. Edinburgh 81A (1978): pp 273-
     286.

11] Campbell, C. M. and Robertson, E. F. "Groups related to $F^{a,b,c}$ involving
     Fibonacci numbers." in The Geometric Vein', edited by C. Davis, B. Grünbaum
     and F. A. Sherk, Springer Verlag (1982): pp 569-576.

12] Campbell, C. M. and Robertson, E. F. "Some problems in group presentations."
     J. Korean Math. Soc. 19 (1983): pp 123-128.

13] Campbell, C. M. and Thomas, R. M. "On infinite groups of Fibonacci type."
     Proc. Edinburgh Math. Soc. 29 (1986): pp 225-232.

14] Chalk, C. P. and Johnson, D. L. "The Fibonacci groups, II." Proc. Roy. Soc.
     Edinburgh 77A (1977): pp 79-86.

15] Conway, J. H. "Advanced problem 5327." Amer. Math. Monthly 72 (1965): p 915.

16] Conway, J. H. et al., "Solution to advanced problem 5327." Amer. Math. Monthly
     74 (1967): pp 91-93.

17] Havas, G. "Computer aided determination of a Fibonacci group." Bull. Austral.
     Math. Soc. 15 (1976) 297-305.

18] Havas, G., Richardson, J. S. and Sterling, L. S. "The last of the Fibonacci
     groups." Proc. Roy. Soc. Edinburgh 83A (1979): pp 199-203.

19] Johnson, D. L. "Some infinite Fibonacci groups." Proc. Edinburgh Math. Soc. 19
     (1975): pp 311-314.

20] Johnson, D. L. "Presentations of groups." L. M. S. Lecture Notes 22, Cambridge
     University Press (1976).

21] Johnson, D. L. "Topics in the theory of group presentations." L. M. S. Lecture
     Notes 42, Cambridge University Press (1980).

22] Johnson, D. L. and Robertson, E. F. "Finite groups of deficiency zero." in
     'Homological group theory', edited by C. T. C. Wall, L. M. S. Lecture Notes 36,
     Cambridge University Press (1979): pp 275-289.

23] Johnson, D. L., Wamsley, J. W. and Wright, D. "The Fibonacci groups." Proc.
     London Math. Soc. 29 (1974): pp 577-592.

24] Lyndon, R. C., unpublished.

[25]   Lyndon, R. C. and Schupp, P. E. "Combinatorial group theory." *Ergebnisse der Mathematik und ihrer Grenzgebiete 89, Springer Verlag* (1977).

[26]   Magnus, W., Karrass, A. and Solitar, D. "Combinatorial group theory." *Dover* (1976).

[27]   Neubuser, J. "An elementary introduction to coset table methods in computational group theory." in *'Groups - St. Andrews 1981', edited by C. M Campbell and E. F. Robertson, L. M. S. Lecture Notes 71, Cambridge University Press* (1982): pp 1-45.

[28]   Rabin, M. O. "Recursive unsolvability of group theoretic problems." *Ann. of Math.* 67 (1958): pp 172-194.

[29]   Rotman, J. J. "The theory of groups." *Allyn and Bacon* (1965).

[30]   Thomas, R. M. "Some infinite Fibonacci groups." *Bull. London Math. Soc.* 15 (1983): pp 384-386.

Shiro Ando

A TRIANGULAR ARRAY WITH HEXAGON PROPERTY, DUAL TO PASCAL'S TRIANGLE

1. INTRODUCTION

Choose any entry A_0 inside Pascal's triangle, and let A_1, A_2, . . . , A_6 be the six terms surrounding it as follows:

$$A_2 \qquad A_3$$

$$A_1 \qquad A_0 \qquad A_4$$

$$A_6 \qquad A_5$$

Hoggatt and Hansell [1] proved the identity

$$A_1 A_3 A_5 = A_2 A_4 A_6, \tag{1}$$

which has been generalized to many other figures in Pascal's triangle and also to the case of multinomial coefficients by many authors. In [2], Gould observed a remarkable property:

$$\gcd (A_1, A_3, A_5) = \gcd (A_2, A_4, A_6) \tag{2}$$

and conjectured that it would be valid even in more general cases, which was established by Hillman and Hoggatt [3]. He also showed that the equality

$$\operatorname{lcm} (A_1, A_3, A_5) = \operatorname{lcm} (A_2, A_4, A_6) \tag{3}$$

61

N. Philippou et al. (eds.), Applications of Fibonacci Numbers, 61–67.
1988 by Kluwer Academic Publishers.

does not always hold.

We will study here similar arrays of numbers in which the roles of greates common divisor and least common multiple will be interchanged.

2. NOTATIONS

Let p be a prime number. For any rational number $r \neq 0$, there exists unique integer v such that

$$r = p^v b/a$$

where both a and b are integers not divisible by p. This v is called the exponentia valuation of r belonging to p, or simply, p-adic valuation of r, and is denoted by $v_p(r)$, or $v(r)$ if there is no chance of confusion. If r is an integer divisible by p then v is the largest positive integer such that $p^v \mid r$.

This valuation satisfies

   (i) $v(1) = 0$,

   (ii) $v(rs) = v(r)+v(s)$,

   (iii) $v(r \pm s) \geq \min(v(r), v(s))$, where equality holds if $v(r) \neq v(s)$ for any rational number r and s.

If we put $v(A_i) = e_i$, then the equalities (1) and (3) are equivalent to having

$$e_1+e_3+e_5 = e_2+e_4+e_6, \tag{4}$$

and

$$\max(e_1, e_3, e_5) = \max(e_2, e_4, e_6) \tag{5}$$

for all prime numbers p, respectively.

3. MAIN RESULTS

First, we consider the triangular array that has

$$(n+1)\binom{n}{k} = (n+1) \, ! \, / \, k \, ! \, \delta \, ! \quad (\, k+\delta=n \,)$$

s its general term instead of $\binom{n}{k}$ of Pascal's triangle. This case is simple and essential, although it is just a special case of theorem 2. We call it the modified Pascal's triangle:

Modified Pascal's Triangle

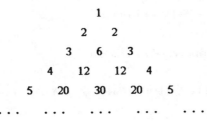

```
            1
         2     2
      3     6     3
   4     12    12    4
5    20    30    20    5
  . . .   . . .   . . .   . . .   . . .
```

Theorem 1: Let

$$A_1 = (n+1) \, ! \, / \, (k-1) \, ! \, (\delta+1) \, !, \quad A_2 = n \, ! \, / \, (k-1) \, ! \, \delta \, !, \quad A_3 = n \, ! \, / \, k \, ! \, (\delta-1) \, !,$$
$$A_4 = (n+1) \, ! \, / \, (k+1) \, ! \, (\delta-1) \, !, \quad A_5 = (n+2) \, ! \, / \, (k+1) \, ! \, \delta \, !,$$
$$A_6 = (n+2) \, ! \, / \, k \, ! \, (\delta+1) \, !$$

be the six terms surrounding an inside entry $A_0 = (n+1) \, ! \, / \, k \, ! \, \delta \, !$ in the modified Pascal's triangle. Then we have

$$A_1 A_3 A_5 = A_2 A_4 A_6, \tag{1}$$

and

$$\text{lcm} \, (A_1, \, A_3, \, A_5) = \text{lcm} \, (A_2, \, A_4, \, A_6), \tag{3}$$

but the equality

$$\gcd \, (A_1, \, A_3, \, A_5) = \gcd \, (A_2, \, A_4, \, A_6) \tag{2}$$

does not always hold.

Proof: (1) is clear. So, we have (4) . In order to prove (5) for all primes p, let us fix any p, and take the p-adic valuation of the equalities

$$(\delta+1)\ A_1 = (n+1)\ A_2 = kA_0\ ,$$
$$(n+1)\ A_3 = (k+1)\ A_4 = \delta A_0\ ,$$
$$(k+1)\ A_5 = (\delta+1)\ A_6 = (n+2)\ A_0\ .$$

The first line of (6) gives

$$v(\delta+1) + e_1 = v\ (n+1) + e_2 = v\ (k) + e_0\ .$$

Using $k = (n+1) - (\delta+1)$, we have

$$v(k) \geq \min\ (v(n+1)\ ,\ v\ (\delta+1))\ ,$$

where the equality holds if $v\ (n+1) \neq v\ (\delta+1)$. From (7) and (8), we have

$$\max\ (e_1,\ e_2) \geq e_0.$$

From the second and the third line of (6), we have also

$$\max\ (e_3,\ e_4) \geq e_0,\ \text{and}\ \max\ (e_5,\ e_6) \geq e_0.$$

In (9) and (10), the equality holds if the two members in the parenthesis on the left hand side are not equal, which means, for instance, $e_1 = e_2$ if $e_1 > e_0$, and so on.

Therefore, if one of the three numbers on the left hand side of (9) and (10) is larger than e_0, then (5) holds. On the other hand, if all of them are equal to e_0, then we can easily conclude from (4) that there is a number equal to e_0 both in e_1, e_3, e_5 and in e_2, e_4, e_6, which leads to the same conclusion.

As a counter example of the greatest common diviser equality, for $k=1$, $\delta=2$, $n=3$, we have $A_1 = 4$, $A_2 = 3$, $A_3 = 6$, $A_4 = 12$, $A_5 = 30$, $A_6 = 20$, where gcd $(A_1, A_3, A_5) = 2$, but gcd $(A_2, A_4, A_6) = 1$.

Next, we define the modified generalized Pascal's triangle, where we replace the sequence of positive integers with other sequences. Let a_1, a_2, a_3, . . . , be sequence of positive integers satisfying

$$\gcd(a_m, a_n) \mid a_{m+n}, \quad \gcd(a_m, a_n) \mid a_{|m-n|}. \tag{11}$$

'hen we replace n by a_n in the modified Pascal's triangle, to form the modified eneralized Pascal's triangle with general term analogous to $\binom{n}{k}$ in Pascal's triangle, iven by

$$\frac{a_1 a_2 \ldots a_{n+1}}{(a_1 a_2 \ldots a_k)(a_1 a_2 \ldots a_\delta)}, \qquad (k+\delta = n).$$

Theorem 2: In the modified generalized Pascal's triangle, for the six terms surrounding an inside entry A_0, we have

$$A_1 A_3 A_5 = A_2 A_4 A_6 \tag{1}$$

and

$$\text{lcm}(A_1, A_3, A_5) = \text{lcm}(A_2, A_4, A_6). \tag{3}$$

Proof: In this case, we can use

$$\left. \begin{array}{l} a_{\delta+1} A_1 = a_{n+1} A_2 = a_k A_0, \\ a_{n+1} A_3 = a_{k+1} A_4 = a_\delta A_0, \\ a_{k+1} A_5 = a_{\delta+1} A_6 = a_{n+2} A_0, \end{array} \right\} \tag{12}$$

instead of (6) in the proof of theorem 1.

From the condition (11), we have inequalities

$$\min(v(a_m), v(a_n)) \le v(a_{m+n}), \tag{13}$$
$$\min(v(a_m), v(a_n)) \le v(a_{|m-n|}). \tag{14}$$

This latter inequality leads to

$$\min(v(a_m), v(a_{m+n})) \le v(a_n),$$

from which we conclude that equality holds in (13) if $v(a_m) > v(a_n)$.

In a similar manner, equality holds in (13) and (14) if $v(a_m) \neq v(a_n)$ so that w can derive (9) and (10) with the same equality conditions as in the proof of theore 1 from (12) to complete the proof.

<u>Corollary:</u> In the triangular array which has modified Fibonomial coefficients

$$\frac{F_1 F_2 \cdots F_{n+1}}{F_1 F_2 \cdots F_k F_1 F_2 \cdots F_\delta} , \quad (k+\delta = n)$$

in its entries, we have equalities (1) and (3), while equality (2) does not always hold.

<u>Proof:</u> Because of the well-known equality:

$$\gcd (F_m, F_n) = F_{(m, n)},$$

where (m, n) stands for the greatest common divisor of m and n, condition (11) in theorem 2 must be satisfied so that we have (1) and (3).

A counter example of equality (2) will be given as follows: let k=2, δ=3, n= then we have $A_1 = 40$, $A_2 = 15$, $A_3 = 30$, $A_4 = 120$, $A_5 = 780$, $A_6 = 520$. Thus, gcd $(A_1, A_3, A_5) = 10$, gcd $(A_2, A_4, A_6) = 5$.

4. REMARKS

Many other properties concerning the greatest common divisor and leas common multiple which hold in Pascal's triangle have their counterparts in ou modified array, where the roles of the greatest common divisor and the leas common multiple are interchanged.

We may consider a more general triangular array which has entries (n+r) ! k ! δ ! (k+δ = n, r is a given non-negative integer), where the equality (1) holds. $r \geq 2$, however, we have neither (2) nor (3) in this generalized array.

ACKNOWLEDGMENTS

I am indebted to Dr. Daihachiro Sato for drawing my attention to this subject, and suggesting various generalizations.

REFERENCES

[1] Hoggatt, Jr., V. E. and Hansell, W. "The Hidden Hexagon Squares." *Fibonacci Quarterly, Vol. 9, No. 2* (1971) p 120, p 133.

[2] Gould, H. W. "A New Greatest Common Divisor Property of the Binomial Coefficients. " *Fibonacci Quarterly, Vol. 10, No. 6* (1972) pp 579-584, p 628.

[3] Hillman, A. F. and Hoggatt, Jr., V. E. "A Proof of Gould's Pascal Hexagon Conjecture." *Fibonacci Quarterly, Vol. 10, No. 6* (1972) pp 565-568, p 598.

Odoardo Brugia and Piero Filipponi

FUNCTIONS OF THE KRONECKER SQUARE OF THE MATRIX Q

1. INTRODUCTION

As for the well-known matrix Q, [1], a number of matrices can be defined so that their successive powers contain entries related to certain Fibonacci numbers.

In this paper we consider a four-by-four matrix, which will be called matrix S, defined as the Kronecker square QxQ. After outlining some properties of S (sec. 2), closed expressions for the entries of any function f(S) are derived on the basis of a polynomial representation of a function g(Q) such that $f(xy) = g(x)g(y)$ for x and y arbitrary quantities (sec. 3). These expressions are used in sec. 4 to evaluate certain series involving squares of Fibonacci and Lucas numbers.

Throughout the paper, as usual, F_k and L_k denote the k-th Fibonacci and Lucas number, respectively, while $\alpha = (1+\sqrt{5})/2$ and $\beta = (1-\sqrt{5})/2$ are the roots of the equation $x^2 - x - 1 = 0$.

2. DEFINITION AND PROPERTIES OF THE MATRIX S.

We recall [2] that, for given matrices $M = [m_{ij}]$ and N of order s and t, respectively, the matrix

$$
\begin{bmatrix}
m_{11}N & m_{12}N & \cdots & m_{1s}N \\
m_{12}N & m_{22}N & \cdots & m_{2s}N \\
\cdots & \cdots & \cdots & \cdots \\
m_{s1}N & m_{s2}N & \cdots & m_{ss}N
\end{bmatrix}
$$

A. N. Philippou et al. (eds.), Applications of Fibonacci Numbers, 69–76.

of order st is called the *Kronecker product* of M and N and written MxN; MxM is called the *Kronecker square* of M. Moreover, we recall [2] that the eigenvalues of MxN are $\mu_i \nu_j$ (i=1,2,. . . ,s; j=1,2,. . . ,t) where μ_i are the eigenvalues of M and ν_j those of N.

Let us define the matrix S as the Kronecker square of the matrix Q

$$S = QxQ = \begin{bmatrix} 1 & 1 \\ 1 & 0 \end{bmatrix} x \begin{bmatrix} 1 & 1 \\ 1 & 0 \end{bmatrix} = \begin{bmatrix} 1 & 1 & 1 & 1 \\ 1 & 0 & 1 & 0 \\ 1 & 1 & 0 & 0 \\ 1 & 0 & 0 & 0 \end{bmatrix}. \tag{1}$$

The matrix S is symmetric and its eigenvalues are

$$\begin{cases} \lambda_1 = \alpha^2 \\ \lambda_2 = \beta^2 \\ \lambda_3 = \lambda_4 = -1, \text{ (multiplicity 1 on the minimal polynomial),} \end{cases} \tag{2}$$

as the eigenvalues of Q are α and β.

Since, for k a nonnegative integer,

$$Q^k = \begin{bmatrix} F_{k+1} & F_k \\ F_k & F_{k-1} \end{bmatrix}$$

and [2] $(QxQ)^k = Q^k x Q^k$, it is readily seen that

$$S^k = \begin{bmatrix} F_{k+1}^2 & F_k F_{k+1} & F_k F_{k+1} & F_k^2 \\ F_k F_{k+1} & F_{k-1} F_{k+1} & F_k^2 & F_k F_{k-1} \\ F_k F_{k+1} & F_k^2 & F_{k-1} F_{k+1} & F_k F_{k-1} \\ F_k^2 & F_k F_{k-1} & F_k F_{k-1} & F_{k-1}^2 \end{bmatrix}. \tag{3}$$

3. POLYNOMIAL REPRESENTATION OF ANY FUNCTION OF S

It is known [3,4] that if f is a generic function defined on the spectrum of a diagonable matrix X with n distinct eigenvalues ξ_j, $(j = 1, 2, \ldots, n)$, then $f(X)$ can be represented as a polynomial in X of degree less than or equal to n-1

$$f(X) = \sum_{i=0}^{n-1} b_i X^i \tag{4}$$

where the coefficients b_i are given by the solution of the system of n equations and n unknowns

$$\sum_{i=0}^{n-1} b_i \xi_j^i = f(\xi_j), \quad (j=1, 2, \ldots, n). \tag{5}$$

In this section a closed form expression for the entries a_{ij}, $(i, j=1, \ldots, 4)$ of the polynomial representation of $f(S)$ is worked out by means of a method simpler than the above mentioned one. Namely, a method is used where only the calculation of a function $g(Q)$ and of the product $g(Q) \times g(Q)$ is required. In this way, a large amount of manipulations involving the use of the Binet's form for F_k and L_k and several Fibonacci identities is avoided.

3.1. DETERMINATION OF g(Q)

Letting $n = 2$, $\xi_1 = \alpha$ and $\xi_2 = \beta$, $g(Q)$ can be derived from (4) and (5). In fact we obtain

$$\begin{cases} b_0 = (\alpha g(\beta) - \beta g(\alpha)) / \sqrt{5} \\ b_1 = (g(\alpha) - g(\beta)) / \sqrt{5} \end{cases} \tag{6}$$

and

$$g(Q) = b_0 I + b_1 Q = \frac{1}{\sqrt{5}} \begin{bmatrix} \alpha g(\alpha) - \beta g(\beta) & g(\alpha) - g(\beta) \\ g(\alpha) - g(\beta) & \alpha g(\beta) - \beta g(\alpha) \end{bmatrix}, \tag{7}$$

where I denotes the two-by-two identity matrix.

3.2. DETERMINATION OF f(S)

Using (7), (1), the identity $Q + I = Q^2$ and some well-known properties [2,p.228] of the Kronecker product of matrices, we can write

$$
\begin{aligned}
g(Q) \times g(Q) &= (b_0 I + b_1 Q) \times (b_0 I + b_1 Q) \\
&= b_0^2 I \times I + b_0 b_1 (I \times Q + Q \times I) + b_1^2 Q \times Q \\
&= b_0^2 I + b_0 b_1 ((I + Q) \times (I + Q) - Q \times Q - I) + b_1^2 Q \times Q \\
&= b_0^2 I + b_0 b_1 (Q^2 \times Q^2 - S - I) + b_1^2 S \\
&= b_0^2 I + b_0 b_1 (S^2 - S - I) + b_1^2 S \\
&= b_0 (b_0 - b_1) I + b_1 (b_1 - b_0) S + b_0 b_1 S^2,
\end{aligned}
\tag{8}
$$

I being the identity matrix of appropriate order.

Since S has three distinct eigenvalues, from (8) and (4) it is apparent that g(Q)xg(Q) can be seen as the polynomial representation $c_0 I + c_1 S + c_2 S^2$ of the function f(S), where

$$
\begin{cases}
c_0 = b_0 (b_0 - b_1) = (\beta^3 g^2(\alpha) + \alpha^3 g^2(\beta) + g(\alpha) g(\beta)) / 5 \\
c_1 = b_1 (b_1 - b_0) = (-\beta^2 g^2(\alpha) - \alpha^2 g^2(\beta) + 3g(\alpha) g(\beta)) / 5 \\
c_2 = b_0 b_1 = (-\beta g^2(\alpha) - \alpha g^2(\beta) + g(\alpha) g(\beta)) / 5 ,
\end{cases}
\tag{9}
$$

in virtue of (6). On the other hand, on the basis of (5) the coefficients c_0, c_1 and c_2 could be expressed in terms of $f(\alpha^2)$, $f(\beta^2)$ and $f(-1)$. Since the equalities (9) hold independently of g, and hence of f, the equalities

$$
\begin{cases}
f(\alpha^2) = g^2(\alpha) \\
f(\beta^2) = g^2(\beta) \\
f(-1) = g(\alpha) g(\beta)
\end{cases}
\tag{10}
$$

must necessarily hold. They are satisfied if $f(xy) = g(x)g(y)$, x and y being arbitrary quantities. It follows that the entries a_{ij} of f(S) can be found carrying out the Kronecker square g(Q)xg(Q) and replacing the right hand sides of (10) by the corresponding left hand sides. In fact, from (7) and (10) we get

$$a_{11} = (\alpha^2 f_1 + \beta^2 f_2 + 2f_3) / 5$$

$$a_{12} = a_{13} = a_{21} = a_{31} = (\alpha f_1 + \beta f_2 - f_3) / 5$$

$$a_{14} = a_{23} = a_{32} = a_{41} = (f_1 + f_2 - 2f_3) / 5$$

$$a_{22} = a_{33} = (f_1 + f_2 + 3f_3) / 5 \tag{11}$$

$$a_{24} = a_{34} = a_{42} = a_{43} = (-\beta f_1 - \alpha f_2 + f_3) / 5$$

$$a_{44} = (\beta^2 f_1 + \alpha^2 f_2 + 2f_3) / 5$$

where

$$f_i = f(\lambda_i), \ (i = 1, 2, 3). \tag{12}$$

4. APPLICATION EXAMPLES

Let us suppose to have at our disposal a matrix X whose successive powers contain entries which are related to certain Fibonacci numbers. Whenever there exists a function f such that f(X) can be obtained by means of two different algorithms yielding $f(X) = A = [a_{ij}]$ and $f(X) = \hat{A} = [\hat{a}_{ij}]$, respectively, it may happen that A and $\hat{A}$ differ formally. In this case equating a_{ij} and $\hat{a}_{ij}$, for some i and j , some Fibonacci identities can be obtained. In the previous paper [5] some identities were worked out using Q as X. In this section we show how using an alternative algorithm to find certain functions of S, some series of Fibonacci and Lucas squares can be evaluated.

4.1. THE EXPONENTIAL OF xS^k

For x an arbitrary real quantity and k a nonnegative integer, xS^k is obviously a polynomial in S. Therefore, the entries a_{ij} of exp (xS^k) can be obtained from (11) specializing f_i to exp $(x\lambda_i^k)$, (i = 1, 2, 3). It must be noted that k = 0 implies that xS^k has not three distinct eigenvalues and (11) does not apply.

On the other hand, exp (xS^k) can be calculated by means of the power series expansion [3]

$$\exp (xS^k) = \sum_{n=0}^{\infty} \frac{x^n S^{kn}}{n!} = [\hat{a}_{ij}].$$

Equating $\hat{a}_{ij}$ and a_{ij}, from (3) and (11) we obtain

$$\sum_{n=0}^{\infty} \frac{x^n F_{kn}^2}{n!} = \frac{\exp(x\alpha^{2k}) + \exp(x\beta^{2k}) - 2\exp(x(-1)^k)}{5} , \tag{12}$$

$$\sum_{n=0}^{\infty} \frac{x^n F_{kn+1}^2}{n!} = \frac{\alpha^2\exp(x\alpha^{2k}) + \beta^2\exp(x\beta^{2k}) + 2\exp(x(-1)^k)}{5} , \tag{13}$$

$$\sum_{n=0}^{\infty} \frac{x^n F_{kn-1}^2}{n!} = \frac{\beta^2\exp(x\alpha^{2k}) + \alpha^2\exp(x\beta^{2k}) + 2\exp(x(-1)^k)}{5} , \tag{14}$$

$$\sum_{n=0}^{\infty} \frac{x^n F_{kn-1}F_{kn+1}}{n!} = \frac{\exp(x\alpha^{2k}) + \exp(x\beta^{2k}) + 3\exp(x(-1)^k)}{5} . \tag{15}$$

Combining (13), (14) and (15) we get

$$\sum_{n=0}^{\infty} \frac{x^n L_{kn}^2}{n!} = \exp(x\alpha^{2k}) + \exp(x\beta^{2k}) + 2\exp(x(-1)^k). \tag{16}$$

Analogous results can be obtained using circular and hyperbolic functions of xS^k .

4.2 THE INVERSE OF $I-xS^k$

Provided

$$x \neq \begin{cases} \alpha^{-2k} \\ \beta^{-2k} \\ (-1)^k , \end{cases}$$

the matris $I-xS^k$ admits its inverse. For $k > 0$, $(I-xS^k)^{-1}$ can be obtained from (11) specializing f_i to $(1-x\lambda_i^k)^{-1}$. On the other hand, provided the inequalities

$$-\alpha^{-2k} < x < \alpha^{-2k}, (k=1, 2, \ldots) \tag{17}$$

hold, $(I-xS^k)^{-1}$ can be calculated by means of the power series expansion [3]

$$(I-xS^k)^{-1} = \sum_{n=0}^{\infty} x^n S^{kn} = [\hat{a}_{ij}].$$

Equating $\hat{a}_{ij}$ and a_{ij}, from (3) and (11), after some algebraic steps we obtain, under the restrictions (17),

$$\sum_{n=0}^{\infty} x^n F_{kn}^2 = \{F_k^2 x + (-1)^k F_k^2 x^2\} / \Delta \tag{18}$$

$$\sum_{n=0}^{\infty} x^n F_{kn+1}^2 = \{1 + (F_{k+1}^2 - L_k^2 + (-1)^k)x + (-1)^k F_{k-1}^2 x^2\} / \Delta \tag{19}$$

$$\sum_{n=0}^{\infty} x^n F_{kn-1}^2 = \{1 + (F_{k-1}^2 - L_k^2 + (-1)^k x + (-1)^k F_{k+1}^2 x^2\} / \Delta \tag{20}$$

$$\sum_{n=0}^{\infty} x^n F_{kn-1} F_{kn+1} = \{1-2(2F_k^2 + (-1)^k)x + (-1)^k F_{k-1}F_{k+1}x^2\} / \Delta, \tag{21}$$

where

$$\Delta = 1 - (L_{2k} + (-1)^k)x + ((-1)^k L_{2k} + 1)x^2 - (-1)^k x^3. \tag{22}$$

Combining (19), (20) and (21) we get

$$\sum_{n=0}^{\infty} x^n L_{kn}^2 = \{4 - (3L_{2k} + 2(-1)^k)x + (-1)^k L_k^2 x^2\} / \Delta. \tag{23}$$

4.3 REMARK

The authors realize that the above results may not be new and that they can be readily obtained in other ways.

In spite of this fact, it is hoped that the idea offered in sec. 4. is fruitful. In other words, it is hoped that the use of suitable Fibonacci matrices together with appropriate functions may lead to the discovery of new Fibonacci identities.

REFERENCES

[1] Hoggatt, Jr., V. E. "Fibonacci and Lucas Numbers." *Houghton Mifflin Co., Boston* (1969).

[2] Bellman, R. "Introduction to Matrix Analysis." *McGraw-Hill Book Co. Inc., New York* (1960).

[3] Gantmacher, F. R. "The Theory of Matrices." *Vol. 1, Chelsea Publ. Co., New York* (1960).

[4] Filipponi, P. and Pesamosca, G. "Calcolo Numerico di Funzioni di Matrice a Partire da una Loro Rappresentazione Polinomiale." *Int. Rept. 3A0778, Fdn. Ugo Bordoni, Roma* (1978).

[5] Filipponi, P. and Horadam, A. F. "A Matrix Approach to Certain Identities." *Int. Rept. 2C4885, Fdn. Ugo Bordoni, Roma* (1985). To Appear in The Fibonacci Quarterly.

Neville Robbins

FIBONACCI NUMBERS OF THE FORMS PX²±1, PX³±1, WHERE P IS PRIME

INTRODUCTION

Let m denote a non-negative integer, F_m the mth Fibonacci number, p a prime. ibonacci numbers of the forms x^2, $2x^2$, x^2+1, x^2-1, px^2, x^3, $2x^3$, px^3, p^2x^3, $x^3\pm1$ have een studied by J. H. E. Cohn [1], R. Finkelstein [3], H. C. Williams [9], the author I, [8], A. Petho [6], H. London & R. Finkelstein [5], and J. C. Lagarias & D. P. eisser [4]. In this article, we find all solutions to each of the four equations:

.) $$F_m = px^2 + 1$$

•) $$F_m = px^2 - 1$$

:) $$F_m = px^3 + 1$$

)) $$F_m = px^3 - 1$$

If $m = 4n+k$, where k is successively ±1, 0, 2, then by using appropriate entities, each of (A) through (D) transforms into

:) $$F_{2n+i}L_{2n+j} = px^c$$

here L_n denotes the nth Lucas number, c = 2 or 3, and i, j are functions of k, with $|i| \le 2$, $|j| \le 2$, $|i-j| = 1$ or 2 in each case. If $(F_{2n+i}, L_{2n+j}) = 1$, which occurs iless i and j are both even, then either F_{2n+i} or $L_{2n+j} = y^c$. In these cases, n, i, j, d hence m can be determined from known results. Otherwise $(F_{2n+i}, L_{2n+j}) = 3$. his leads to a more complicated situation, whose resolution depends in part on emmas 1 through 5 below.

As a byproduct, we obtain all Fibonacci numbers of the forms $p\pm1$, namely: 1, 3, and 8.

77

N. Philippou et al. (eds.), Applications of Fibonacci Numbers, 77–88.
1988 by Kluwer Academic Publishers.

PRELIMINARIES

$F_n = x^2$ iff $n = 0, 1, 2, 12$ (

$F_n = 2x^2$ iff $n = 0, 3, 6$ (

$L_n = x^2$ iff $n = 1, 3$ (

$L_n = 2x^2$ iff $n = 0, 6$ (

$F_{2n} = F_n L_n$ (

$L_{3n} = L_n(L_n^2 - 3(-1)^n)$ (

$2 \mid L_n$ iff $3 \mid n$ (

$3 \mid L_n$ iff $4 \mid (n-2)$ (

$3^k \mid F_n$ iff $4(3^{k-1}) \mid n$ (

$2 \mid F_n$ iff $3 \mid n$ (1

If $k \geq 3$, then $2^k \mid F_n$ iff $3(2^{k-2}) \mid n$ (1

$(F_n, L_n) = \begin{cases} 2 \text{ if } 3 \mid n \\ 1 \text{ if } 3 \nmid n \end{cases}$ (1

$F_m \mid F_n$ iff $m \mid n$ (1

$F_{4n \pm 1} = F_{2n} L_{2n \pm 1} + 1$ (1

$F_{4n} = F_{2n+1} L_{2n-1} + 1$ (1

$F_{4n-2} = F_{2n-2} L_{2n} + 1$ (1

$F_{4n \pm 1} = F_{2n \pm 1} L_{2n} - 1$ (1

$F_{4n} = F_{2n-1} L_{2n+1} - 1$ (1

$F_{4n-2} = F_{2n} L_{2n-2} - 1$ (1

$(F_m, L_{m \pm n}) \mid L_n$ (2

$(F_n, L_{n \pm 1}) = 1$ (2

$(F_{2n \pm 1}, L_{2n \mp 1}) = 1$ (2

$F_n = 3x^2$ iff $n = 0, 4$ (2

$F_{2n} = px^2$ iff $2n = 4, p = 3$ (2

$L_n = x^3$ iff $n = 1$ (2

$L_n = 2x^3$ iff $n = 0$ (2

$F_n = x^3$ iff $n = 0, 1, 2, 6$ (2

$F_n = 2x^3$ iff $n = 0, 3$ (2

$F_n = 3x^3$ iff $n = 0, 4$ (2

$F_n = 4x^3$ iff $n = 0$ (30

$$F_n = 9x^3 \quad \text{iff} \quad n = 0 \tag{31}$$

$$4 \mid L_n \quad \text{iff} \quad n \equiv 3 \quad (\text{mod } 6) \tag{32}$$

$$9 \mid L_n \quad \text{iff} \quad n \equiv 6 \quad (\text{mod } 12) \tag{33}$$

Remarks: (1) through (4) appear in [1]; (5) through (13) as well as (32) and (33) are well-known; (14) through (19) appear in [2]; (20) through (22) appear in [7]. (23) and (24) follow from Theorem 1 in [8]. (25) through (31) follow from Theorems 2 and 5 in [4].

Lemma 1: (I) $F_n = 6x^2$ iff $n = 0$

Proof: It suffices to show that (I) has no solution for $n > 0$. Assume that such a solution exists. Then (9) and (10) imply $12 \mid n$. Now (13) implies $F_{12} \mid F_n$, that is $144 \mid 6x^2$, so $24 \mid x^2$. Since $2 \mid x$, let $x = 2y$, so $6 \mid y^2$. Therefore $6 \mid y$, so $36 \mid y^2$, $144 \mid x^2$, and $864 \mid 6x^2$, that is, $2^5 3^3 \mid F_n$. Now (9) and (11) imply $72 \mid n$. Let $n = 2m$, where $36 \mid m$, so $m \geq 36$. Now hypothesis and (5) imply $F_m L_m = 6x^2$, so (12) implies $(\frac{1}{2}F_m)(\frac{1}{2}L_m) = 6y^2$, also $(\frac{1}{2}F_m, \frac{1}{2}L_m) = 1$. Therefore one of the following must hold:

(a) $\quad F_m = 2s^2, \quad L_m = 12t^2$

(b) $\quad F_m = 4s^2, \quad L_m = 6t^2$

(c) $\quad F_m = 6s^2, \quad L_m = 4t^2$

(d) $\quad F_m = 12s^2, \quad L_m = 2t^2$

But then (1) through (4) imply $m \leq 12$, an impossibility.

Lemma 2: (II) $\quad L_n = 6x^2$ is impossible.

Proof: If (II) holds, then (8) and (7) imply $n = 3m$, where $4 \mid (m-2)$. By hypothesis and (6), we have $L_m(L_m^2-3) = 6x^2$. Now (8) implies $(L_m, L_m^2-3) = 3$. Therefore $(L_m/3, (L_m^2-3)/3) = 1$. We have $(L_m/3)((L_m^2-3)/3) = 6(x/3)^2$. Therefore $L_m/3 = as^2$, $(L_m^2-3)/3 = bt^2$, where $ab = 6$. Now $L_m \equiv 0 \pmod 3$ implies $(L_m^2-3)/3 \equiv 2 \pmod 3$, so $bt^2 \equiv 2 \pmod 3$. Since $b \mid 6$, we must have $b = 2$, hence $a = 3$. But then $L_m = (3s)^2$, which contradicts (3).

Lemma 3: The system :   (III) $F_{4k} = 3py^2$, $L_{4k\pm2} = 3z^2$ admits only the trivial
solution: k = 0.

Proof: $F_0 = 0 = 3p(0)^2$, $L_{\pm2} = 3 = 3(1)^2$, so k = 0 is a solution of (III). If (III) has
a solution with k > 0, then p $\neq$ 3, otherwise (1) implies k = 3, so $3z^2 = L_{10}$
or L_{14}, that is $3z^2$ = 123 or 843, an impossiblity. (5) implies $F_{2k}L_{2k} = 3py^2$.
If $3 \nmid k$, then (12) implies $(F_{2k}, L_{2k}) = 1$, so one of the following must hold

$$\begin{array}{lll} \text{(a)} & F_{2k} = s^2, & L_{2k} = 3pt^2 \\ \text{(b)} & F_{2k} = 3s^2, & L_{2k} = pt^2 \\ \text{(c)} & F_{2k} = ps^2, & L_{2k} = 3t^2 \\ \text{(d)} & F_{2k} = 3ps^2, & L_{2k} = t^2 \end{array}$$

If (a), then (1) implies 2k = 2, so $L_{2k} = L_2 = 3 = 3pt^2$, an impossibility; if
(b), then (24) implies 2k = 4, so $3z^2 = L_6$ or L_{10}, that is, $3z^2$ = 18 or 123, an
impossibility; if (c), then (24) implies p = 3, a contradiction; if (d), then (3)
implies 2k = 1 or 3, an impossibility.

If 3 | k, then (12) implies $(F_{2k}, L_{2k}) = 2$, so $(\frac{1}{2}F_{2k}, \frac{1}{2}L_{2k}) = 1$. Also $(\frac{1}{2}F_{2k})$
$(\frac{1}{2}L_{2k}) = 3p(\frac{1}{2}y)^2$, = so one of the following must hold:

$$\begin{array}{lll} \text{(e)} & F_{2k} = 2s^2, & L_{2k} = 6pt^2 \\ \text{(f)} & F_{2k} = 6s^2, & L_{2k} = 2pt^2 \\ \text{(g)} & F_{2k} = 2ps^2, & L_{2k} = 12t^2 \\ \text{(h)} & F_{2k} = 6ps^2, & L_{2k} = 2t^2 \end{array}$$

If (e), then (2) implies 2k = 6, so $L_{2k} = L_6 = 18 = 6pt^2$, so p = 3, a
contradiction; Lemma 1 implies (f) is impossible; Lemma 2 implies (g) is
impossible; if (h), then (4) implies 2k = 6, but (9) implies 4 | 2k, a
contradiction.

Theorem 1: If m > 0, then (A) $F_m = px^2 + 1$ iff m = 1, 2, 4, 6, 7, 8, 23, 25.

Proof:

Case 1: Let $m = 4n+1$. Hypothesis and (14) imply $F_{2n}L_{2n+1} = px^2$. Now (21) implies either (i) $F_{2n} = y^2$, $L_{2n+1} = pz^2$, or (ii) $F_{2n} = py^2$, $L_{2n+1} = z^2$. If (i), then (1) implies $2n = 0$, 2, or 12, so $2n+1 = 1$, 3, or 13. Now $L_1 = 1 \neq pz^2$; $L_3 = 4 \neq pz^2$, but $L_{13} = 521 = 521(1)^2$. Therefore $m = 25$, and $F_{26} = 75025 = 521(12)^2+1$. If (ii), then (3) implies $2n+1 = 1$ or 3, so $2n = 0$ or 2. Now $F_2 = 1 \neq py^2$, but $F_0 = 0 = p(0)^2$. Therefore $m = 1$, and $F_1 = 1 = p(0)^2+1$, with arbitrary p.

Case 2: Let $m = 4n-1$. Hypothesis and (14) imply $F_{2n}L_{2n-1} = px^2$. Now (21) implies either (i) $F_{2n} = y^2$, $L_{2n-1} = pz^2$, or (ii) $F_{2n} = py^2$, $L_{2n-1} = z^2$. If (i), then (1) implies $2n = 2$ or 12 (since hypothesis implies $n > 0$), so $2n-1 = 1$ or 11. Now $L_1 = 1 \neq pz^2$, but $L_{11} = 199 = 199(1)^2$. Therefore $m = 23$, and $F_{23} = 28657 = 199 (12)^2+1$. If (ii), then (3) implies $2n-1 = 1$ or 3, so $2n = 2$ or 4. Now $F_2 = 1 \neq py^2$, but $F_4 = 3 = 3(1)^2$. Therefore $m = 7$, and $F_7 = 13 = 3(2)^2+1$.

Case 3: Let $m = 4n$. Hypothesis and (15) imply $F_{2n+1}L_{2n-1} = px^2$. (22) implies either (i) $F_{2n+1} = y^2$, $L_{2n-1} = pz^2$, or (ii) $F_{2n+1} = py^2$, $L_{2n-1} = z^2$. If (i), then (1) implies $2n+1 = 1$, so $L_{2n-1} = L_{-1} = -1 = pz^2$, an impossibility. If (ii), then (3) implies $2n-1 = 1$ or 3, so $m = 4$ or 8. $F_4 = 3 = 2(1)^2+1$; $F_8 = 21 = 5(2)^2+1$.

Case 4: Let $m = 4n-2$. Hypothesis and (16) imply $F_{2n-2}L_{2n} = px^2$. Let $d = (F_{2n-2}, L_{2n})$. (20) implies $d = 1$ or 3. If $d = 1$, then (3) implies $F_{2n-2} = y^2$, $L_{2n} = pz^2$. (9) and (8) imply $4 \nmid (2n-2)$. Therefore (1) implies $2n-2 = 2$, so $m = 6$. $F_6 = 8 = 7(1)^2+1$. If $d = 3$, then (9) and (8) imply $4 \mid (2n-2)$, so let $2n-2 = 4k$. Now $(F_{4k}/3, L_{4k+2}/3) = 1$ and $(F_{4k}/3)(L_{4k+2}/3) = p(x/3)^2$, so either (i) $F_{4k} = 3r^2$, $L_{4k+2} = 3ps^2$, or (ii) $F_{4k} = 3pr^2$, $L_{4k+2} = 3s^2$. If (i), then (23) implies $4k = 0$ or 4, so $3ps^2 = L_2$ or L_6, that is, $3ps^2 = 3$ or 18, an impossibility. If (ii), then Lemma 3 implies $k = 0$, so $m = 2$. $F_2 = 1 = p(0)+1$, with arbitrary p. We summarize the results in

Table 1

m	1	2	4	6	7	8	23	25
p	arb	arb	2	7	3	5	199	521
x^2	0	0	1	1	4	4	144	144

We have seen that every solution (m, p, x^2) of (A) appears in Table 1. It is easily verified that every entry (m, p, x^2) in Table 1 is a solution of (A). We therefore conclude that all solutions of (A) are given by Table 1.

Corollary 1: $F_m = p+1$ iff $(m, p) = (4, 2)$ or $(6, 7)$.

Proof: Follows from Theorem 1, with $x^2 = 1$.

Theorem 2: If $m > 0$, then (B) $F_m = px^2-1$ iff $m = 1, 2$, or 3.

Proof:

Case 1: Let $m = 4n \pm 1$. Hypothesis and (17) imply $F_{2n\pm1}L_{2n} = px^2$. (20) and (3) imply $F_{2n\pm1} = y^2$, $L_{2n} = pz^2$. Now (1) implies $2n \pm 1 = 1$, so $n = 0$ or 1, and $m = 1$ or 3. $F_1 = 1 = 2(1)^2-1$; $F_3 = 2 = 3(1)^2-1$.

Case 2: Let $m = 4n$. Hypothesis and (18) imply $F_{2n-1}L_{2n+1} = px^2$. (22) implies either (i) $F_{2n-1} = y^2$, $L_{2n+1} = pz^2$, or (ii) $F_{2n-1} = py^2$, $L_{2n+1} = z^2$. If (i), then (1) implies $2n-1 = 1$, so $L_{2n+1} = L_3 = 4 = pz^2$, an impossibility. If (ii), then (3) implies $2n+1 = 1$ or 3, so $F_{2n-1} = F_{\pm1} = 1 = py^2$, an impossibility.

Case 3: Let $m = 4n-2$. Hypothesis and (19) imply $F_{2n}L_{2n-2} = px^2$. Let $d = (F_{2n}, L_{2n-2})$. (20) implies $d = 1$ or 3. If $d = 1$, then (3) implies $F_{2n} = y^2$, $L_{2n-2} = pz^2$. (9) implies $4 \nmid 2n$, so (1) implies $2n = 2$. Therefore $m = 2$, and $F_2 = 1 = 2(1)^2-1$. If $d = 3$, then (9) and (8) imply $4 \mid 2n$, so let $2n = 4k$. Now $(F_{4k}/3,$

$L_{4k-2}/3) = 1$ and $(F_{4k}/3) (L_{4k-2}/3) = p(X/3)^2$, so either (i) $F_{4k} = 3r^2$, $L_{4k-2} = 3ps^2$, or (ii) $F_{4k} = 3pr^2$, $L_{4k-2} = 3s^2$. If (i), then (23) implies $4k = 0$ or 4, so $L_{4k-2} = L_{+2} = 3 = 3ps^2$, an impossiblity. If (ii), then Lemma 3 implies $k = 0$, so $m = -2$, a contradiction.

We summarize the results in

Table 2

m	1	2	3
p	2	2	3
x^2	1	1	1

We have seen that every solution (m, p, x^2) of (B) appears in Table 2. It is easily verified that every entry (m, p, x^2) in Table 2 is a solution of (B). We therefore conclude that all solutions of (B) are given by Table 2.

Corollary 2: $F_m = p-1$ iff (m, p) = (1, 2), (2, 2), or (3, 3)

Proof: Follows from Theorem 2.

Lemma 4: The system (IV) $\begin{cases} F_{4k} = 3py^3 \\ L_{4k+2} = 9z^3 \end{cases}$ has no solution.

Proof: If (IV) has a solution, then (20) implies $p \neq 3$. Hypothesis and (5) imply $F_{2k}L_{2k} = 3py^3$. Now (33) implies $3 \nmid k$, so (12) implies $(F_k, L_k) = (F_{2k}, L_{2k}) = 1$. Therefore one of the following must hold:

(a) $F_{2k} = s^3$, $L_{2k} = 3pt^3$

(b) $F_{2k} = 3s^3$, $L_{2k} = pt^3$

(c) $F_{2k} = ps^3$, $L_{2k} = 3t^3$

(d) $F_{2k} = 3ps^3$, $L_{2k} = t^3$

If (a), then (27) implies $2k = 2$, so $L_{2k} = L_2 = 3pt^3$, an impossibility. If (b), then (29) implies $2k = 4$, so $4k \pm 2 = 6$ or 10. But then L_6 or $L_{10} = 9z^3$, that is, $9z^3 = 18$ or 123, an impossibility. If (c), then (5) and (12) imply F_k or L_k is a cube, so (25) and (27) imply $k = 1$ or 2. If $k = 1$, then $F_{2k} = F_2 = 1 = ps^3$, an impossibility. If $k = 2$, then $L_{2k} = L_4 = 7 = 3t^3$, an impossibility. By (25), (d) is impossible.

__Lemma 5__: The system (V) $\begin{cases} F_{4k} = 9py^3 \\ L_{4k \pm 2} = 3z^3 \end{cases}$ admits only the trivial solution: $k = 0$.

__Proof__: It suffices to show that (V) has no solution for $k > 0$. Suppose that such a solution exists. (27) implies $p \neq 3$. (9) implies $3 \mid k$. Hypothesis and (5) imply $F_{2k}L_{2k} = 9py^3$. (12) implies $(F_{2k}, L_{2k}) = 2$. Therefore $4 \mid 9py^3$, so $2 \mid y^3$, $2 \mid y$. Let $y = 2v$. Therefore $F_{2k}L_{2k} = 72pv^3$. Now (32) implies $4 \mid L_{2k}$; (26) implies $L_{2k} \neq 2x^3$; (30) implies $F_n \neq 4x^3$. Therefore one of the following must hold:

$$\text{(a)} \quad F_{2k} = 36s^3, \quad L_{2k} = 2pt^3$$
$$\text{(b)} \quad F_{2k} = 4ps^3, \quad L_{2k} = 18t^3$$

In either case, (5), (12), and (26) imply $F_k = 2a^3$, $L_{2k} = 18t^3$ or $2pt^3$. Now (28) implies $k = 3$, but then L_{10} or $L_{14} = 3z^3$, that is, 123 or $843 = 3z^3$, an impossibility.

__Theorem 3__: If $m > 0$, then (C) $F_m = px^3 + 1$ iff $m = 1, 2, 4, 6, 10, 11, 13,$ or 14.

__Proof__:

__Case 1__: Let $m = 4n + 1$. Hypothesis and (14) imply $F_{2n}L_{2n+1} = px^3$. Now (21) implies either (i) $F_{2n} = y^3$, $L_{2n+1} = pz^3$, or (ii) $F_{2n} = py^3$, $L_{2n+1} = z^3$. If (i), then (27) implies $2n = 0, 2,$ or 6. Now $L_1 = 1 \neq pz^3$; $L_3 = 4 \neq pz^3$. But $L_7 = 29 = 29(1)^3$. Therefore $m = 13$, and $F_m = F_{13} = 233 = 29(2)^3 + 1$. If (ii), then (25) implies $2n + 1 = 1$, so $m = 1$ and $F_m = F_1 = 1 = p(0)^3 + 1$, with arbitrary p.

Case 2: Let $m = 4n-1$. Hypothesis and (14) imply $F_{2n}L_{2n-1} = px^3$. (21) implies either
(i) $F_{2n} = y^3$, $L_{2n-1} = pz^3$ or (ii) $F_{2n} = py^3$, $L_{2n-1} = z^3$. If (i), then since m>0,
(27) implies $2n = 2$ or 6. Now $L_1 = 1 \neq pz^3$. But $L_5 = 11 = 11(1)^3$.
Therefore $m = 11$, and $F_m = F_{11} = 89 = 11(2)^3+1$. If (ii), then (26) implies
$2n-1 = 1$, so $F_{2n} = F_2 = 1 = py^3$, an impossibility.

Case 3: Let $m = 4n$. Hypothesis and (15) imply $F_{2n+1}L_{2n-1} = px^3$. (22) implies either
(i) $F_{2n+1} = y^3$, $L_{2n-1} = pz^3$, or (ii) $F_{2n+1} = py^3$, $L_{2n-1} = z^3$. If (i), then (27)
implies $2n+1 = 1$, so $L_{2n-1} = L_{-1} = -1 = pz^3$, an impossibility. If (ii), then
(25) implies $2n-1 = 1$, so $n = 1$ and $m = 4$. $F_m = F_4 = 3 = 2(1)^3+1$.

Case 4: Let $m = 4n-2$. Hypothesis and (16) imply $F_{2n-2}L_{2n} = px^3$. Let $d = (F_{2n-2},$
$L_{2n})$. (20) implies $d = 1$ or 3. If $d = 1$, then (8) and (9) imply $4 \nmid (2n-2)$, and
(25) implies $F_{2n-2} = y^3$, $L_{2n} = pz^3$. Now (27) implies $2n-2 = 2$ or 6, so $m = 6$
or 14. Now $F_6 = 8 = 7(1)^3+1$; $F_{14} = 377 = 47(2)^3+1$. If $d = 3$, then (8) and
(9) imply $4 \mid (2n-2)$. Let $2n-2 = 4k$. We have $F_{4k}L_{4k+2} = px^3$. Now $3^2 \mid px^3$,
so $3 \mid x$, that is, $x = 3y$. We have $F_{4k}L_{4k+2} = 27py^3$, so one of the
following must hold;

(a) $F_{4k} = 3s^3$, $L_{4k+2} = 9pt^3$
(b) $F_{4k} = 9s^3$, $L_{4k+2} = 3pt^3$
(c) $F_{4k} = 3ps^3$, $L_{4k+2} = 9t^3$
(d) $F_{4k} = 9ps^3$, $L_{4k+2} = 3t^3$

If (a), then (29) and (33) imply $4k = 4$, so $m = 10$, and $F_m = F_{10} = 55 =$
$2(3)^3+1$. If (b), then (31) implies $k = 0$, so $L_{4k+2} = L_2 = 3 = 3pt^3$, an
impossibility. (c) is impossible by Lemma 4. If (d), then Lemma 5 implies k
$= 0$, so $m = 2$, and $F_m = F_2 = 1 = p(0)^3+1$, with arbitrary p.

We summarize the results in

Table 3

m	1	2	4	6	10	11	13	14
p	arb	arb	2	7	2	11	29	47
x^3	0	0	1	1	27	8	8	8

We have seen that every solution (m, p, x^3) of (C) appears in Table 3. It is easily verified that every entry (m, p, x^3) in Table 3 is a solution of (C). We therefore conclude that all solutions of (C) are given by Table 3.

Theorem 4: If $m \geq 0$, then (D) $F_m = px^3 - 1$ iff $m = 1, 2, 3,$ or 10

Proof:

Case 1: Let $m = 4n \pm 1$. Hypothesis and (17) imply $F_{2n \pm 1} L_{2n} = px^3$. (21) and (25) imply $F_{2n \pm 1} = y^3$, $L_{2n} = pz^3$. (27) implies $2n \pm 1 = 1$, so $n = 0$ or 1, and $m = 1$ or 3. $F_1 = 1 = 2(1)^3 - 1$; $F_3 = 2 = 3(1)^3 - 1$.

Case 2: Let $m = 4n$. Hypothesis and (18) imply $F_{2n-1} L_{2n+1} = px^3$. (22) implies either (i) $F_{2n-1} = y^3$, $L_{2n+1} = pz^3$, or (ii) $F_{2n-1} = py^3$, $L_{2n+1} = z^3$. If (i), then (27) implies $2n-1 = 1$, so $L_{2n+1} = L_3 = 4 = pz^3$, an impossibility. If (ii), then (25) implies $2n+1 = 1$, so $n = m = 0$, so $F_0 = 0 = px^3 - 1$, an impossibility.

Case 3: Let $m = 4n-2$, so $n \geq 1$. Hypothesis and (19) imply $F_{2n} L_{2n-2} = px^3$. Let $d = (F_{2n}, L_{2n-2})$. (20) implies $d = 1$ or 3. If $d = 1$, then (25) implies $F_{2n} = y^3$, $L_{2n-2} = pz^3$. Now (27) implies $2n = 2$ or 6, so $m = 2$ or 10. $F_2 = 1 = 2(1)^3 - 1$; $F_{10} = 55 = 7(2)^3 - 1$. If $d = 3$, then (9) implies $2n = 4k$, and as in the proof of Theorem 3, one of the following must hold:

$$\text{(a)} \quad F_{4k} = 3s^3, \qquad L_{4k-2} = 9pt^3$$
$$\text{(b)} \quad F_{4k} = 9s^3, \qquad L_{4k-2} = 3pt^3$$
$$\text{(c)} \quad F_{4k} = 3ps^3, \qquad L_{4k-2} = 9t^3$$
$$\text{(d)} \quad F_{4k} = 9ps^3, \qquad L_{rk-2} = 3t^3$$

If (a), then (29) and (33) imply $4k = 4$, so $L_{4k-2} = L_2 = 3 = 9pt^3$, an impossibility. If (b) then (31) implies $k = n = 0$, an impossibility. (C) is impossible by Lemma 4. If (d), then Lemma 5 implies $k = n = 0$, an impossibility. We summarize the results in

Table 4

m	1	2	3	10
p	2	2	3	7
x^3	1	1	1	8

We have seen that every solution (m, p, x^3) of (D) appears in Table 4. It is easily verified that every entry (m, p, x^3) in Table 4 is a solution of (D). We therefore conclude that all solutions of (D) are given by Table 4.

REFERENCES

[1] Cohn, J. H. E. "Square Fibonacci numbers, etc." *The Fibonacci Quarterly 2*, (1964): pp 109-113.

[2] Dudley, U. & Tucker, B. "Greatest common divisors in altered Fibonacci sequences." *The Fibonacci Quarterly. 9*, (1971): pp 89-92.

[3] Finkelstein, R. "On Fibonacci numbers which are one more than a square." *J. reine u. angew. Math. 262/263*, (1973): pp 171-182.

[4] Lagarias, J. C. and Weisser, D. P. "Fibonacci and Lucas cubes." *The Fibonacci Quarterly 19*, (1981): pp 39-43.

[5] London, H. and Finkelstein, R. "On Fibonacci and Lucas numbers which are perefect powers." *The Fibonacci Quarterly 7*, (1969): pp 476-481.

[6] Petho, A. "Full cubes in the Fibonacci sequence." *Publ. Math. Debrecen 30* (1983): pp 117-127.

[7] Robbins, N. "On Fibonacci and Lucas numbers of the forms w^2-1, $w^3\pm1$." *The Fibonacci Quarterly 19*, (1981): pp 369-373.

[8] Robbins, N. "On Fibonacci numbers of the form px^2, where p is prime." *The Fibonacci Quarterly 21*, (1983): pp 266-271.

[9] Williams, H. C. "On Fibonacci numbers of the form k^2+1." *The Fibonacci Quarterly 13*, (1975): pp 213-214.

George N. Philippou

ON THE K-TH ORDER LINEAR RECURRENCE AND SOME PROBABILITY APPLICATIONS

INTRODUCTION AND SUMMARY

Let k be a fixed integer greater than one, and let p and x be real numbers in the intervals $(0, 1)$ and $(0, \infty)$, respectively. Consider the sequence of functions $y_n^{(k)}(x)\}_{n=0}^{\infty}$, defined by the linear recurrence of order k

$$y_{n+k}^{(k)}(x) = \sum_{i=1}^{k} a_i y_{n+k-i}^{(k)}(x), \quad n \geq 0, \tag{1.1}$$

where $y_i^{(k)}(x)$ $(0 \leq i \leq k - 1)$ are specified by the initial conditions, and $a_1, \ldots, a_k$ are coefficients which may depend on x.

Denote by $f_n^{(k)}(x)$, $F_n^{(k)}(x)$ and $S_n^{(k)}(p)$, respectively, the Fibonacci polynomials of order k [6], the Fibonacci-type polynomials of order k [7], and the polynacci polynomials of order k [3, 9]. They all are special cases of (1.1). In fact:

If $y_0^{(k)}(x) = 0$, $y_1^{(k)}(x) = 1$, $y_i^{(k)}(x) = \sum_{j=0}^{i} x^{k-j} y_{i-j}^{(k)}(x)$ $(2 \leq i \leq k-1)$, (1.1a)

and $a_i = x^{k-i}$ $(1 \leq i \leq k)$, then $y_n^{(k)}(x) = f_n^{(k)}(x)$, $n \geq 0$.

If $y_0^{(k)}(x) = 0$, $y_1^{(k)}(x) = 1$, $y_i^{(k)}(x) = x \sum_{j=1}^{i} y_{i-j}^{(k)}(x)$ $(2 \leq i \leq k-1)$, (1.1b)

and $a_i = x$ $(1 \leq i \leq k)$, then $y_n^{(k)}(x) = F_n^{(k)}(x)$, $n \geq 0$.

If $y_0^{(k)}(x) = 0$, $y_1^{(k)}(x) = 1$, $y_i^{(k)}(x) = q \sum_{j=1}^{i} p^{j-1} y_{i-j}^{(k)}(x)$ $(2 \leq i \leq k-1)$, (1.1c)

and $a_i = qp^{i-1}$ $(1 \leq i \leq k)$, then $y_n^{(k)}(p) = S_n^{(k)}(p)$, $n \geq 0$.

In (1.1c), and in the sequel, q = 1-p.

A. N. Philippou et al. (eds.), Applications of Fibonacci Numbers, 89–96.
© 1988 by Kluwer Academic Publishers.

Recently, Philippou, Georgiou and Philippou [6, 7] established closed formulas

for $f_{n+1}^{(k)}(x)$ and $F_{n+1}^{(k)}(x)$ ($n \geq 0$), in terms of the multinomial coefficients, as well as in terms of the binomial coefficients. They also related these polynomials to the number of Bernoulli trials N_k until the occurrence of the kth consecutive success.

Presently, a multinomial expansion is derived for $y_{n+1}^{(k)}(x)$ (see Theorem 2.1) which generalizes the multinomial expansions of [6] and [7]. In addition, the polynacci polynomials of order k are shown to be related to N_k (see Proposition 3.1), and a reccurrence relation for $P(N_k = n)$ [1, 4] is reestablished as a corollary. For a recent study of the linear recurrence of order k, we refere to Levesque [2].

2. MULTINOMIAL EXPANSIONS OF $Y_n^{(k)}(x)$

In this section we employ a result of Stanley [10] and the methodology of [6] to obtain a multinomial expansion of $y_{n+1}^{(k)}(x)$ ($n \geq 0$).

Lemma 2.1 [10]: Let $\{y_n^{(k)}(x)\}_{n=0}^{\infty}$ be the sequence of functions defined by (1.1) and let g(s, x) be its generating function. Then g(s, x) is analytic in the interval $|s| < p \equiv \min \{ |r_1|^{-1}, \ldots, |r_k|^{-1} \}$, where $r_1, \ldots, r_k$ are the roots of the associated characteristic equation, and it is given by

$$g(s, x) = \frac{y_0^{(k)}(x) + \sum_{i=1}^{k-1} s^i \left[y_i^{(k)}(x) - \sum_{j=1}^{i} a_j y_{i-j}^{(k)}(x) \right]}{1 - (a_1 s + a_2 s^2 + \ldots + a_k s^k)} \qquad (2.1)$$

Remark 2.1: Let $g_1(s, x)$, $g_2(s, x)$ and $g_3(s, x)$ be the generating functions, respectively of $\{f_n^{(k)}(x)\}_{n=0}^{\infty}$, $\{F_n^{(k)}(x)\}_{n=0}^{\infty}$ and $\{S_n^{(k)}(x)\}_{n=0}^{\infty}$. Also let p_1, p_2 and p_3, respectively, be the appropriate restriction of p in each of the preceeding cases. Then (2.1) and (1.1a)-(1.1c) imply

$$g_1(s, x) = \frac{s}{1 - x^k \left[s/x + (s/x)^2 + \ldots + (s/x)^k \right]}, \quad |s| < p_1, \qquad (2.2)$$

$$g_2(s, x) = \frac{s}{1 - x(s + s^2 + \ldots + s^k)}, \quad |s| < p_2, \qquad (2.3)$$

and

$$g_3(s, x) = \frac{s}{1 - qs\left[1 + ps + (ps)^2 + \ldots + (ps)^{k-1}\right]} \, , \, | \, s \, | < p_3. \tag{2.4}$$

We proceed now to state and prove the main result of this paper.

<u>Theorem 2.1</u>: Let $\{y_n^{(k)}(x)\}_{n=0}^{\infty}$ be the sequence of functions defined by (1.1), and set

$y_0^{(k)}(x) = 0.$ Then

$$y_{n+1}^{(k)}(x) = \sum_{i=1}^{k-1} u_i^{(k)}(x) \sum_{\substack{n_1, \ldots, n_k}} \binom{n_1 + \ldots + n_k}{n_1, \ldots, n_k} a_1^{n_1} \ldots a_k^{n_k}, \quad n \geq 0. \tag{2.5}$$

$$n_1 + 2n_2 + \ldots + kn_k = n + 1 - i$$

where $u_i^{(k)}(x) = y_i^{(k)}(x) - \sum_{j=1}^{i} a_j y_{i-j}^{(k)}(x) \quad (1 \leq i \leq k - 1).$

Here, and in the sequel, $n_1, \ldots, n_k$ are non-negative integers as specified.

<u>Proof</u>: We employ the methodology of [6]; i.e., we use the generating function of $\{y_n^{(k)}(x)\}_{n=0}^{\infty}$, and apply consecutively the multinomial theorem and the transformation $n_i - m_i \ (1 \leq i \leq k)$ and $n - m - \sum_{i=2}^{k} (i - 1)m_i.$
We get

$$\sum_{n=0}^{\infty} s^n y_n^{(k)}(x) = \sum_{i=1}^{k-1} s^i u_i^{(k)}(x) \sum_{n=0}^{\infty} (a_1 s + a_2 s^2 + \ldots + a_k s^k)^n$$

$$= \sum_{i=1}^{k-1} s^i u_i^{(k)}(x) \sum_{n=0}^{\infty} \sum_{\substack{n_1, \ldots, n_1}} \binom{n}{n_1, \ldots, n_k} a_1^{n_1} \ldots a_k^{n_k} s^{n_1 + 2n_2 + \ldots + kn_k}$$

$$n_1 + \ldots + n_k = n$$

$$- \sum_{i=1}^{k-1} s^i u_i^{(k)}(x) \sum_{m=0}^{\infty} s^m \sum_{\substack{m_1, \ldots, m_k \\ m_1 + 2m_2 + \ldots + km_k = m}} \binom{m_1+\ldots+m_k}{m_1, \ldots, m_k} a_1^{m_1} \ldots a_k^{m_k},$$

from which (2.5) follows.

To Theorem 2.1 we have the following simple corollary, by means of (1.1a) - (1.1c).

Corollary 2.1: Let $\{f_n^{(k)}(x)\}_{n=0}^{\infty}$, $\{F_n^{(k)}(x)\}_{n=0}^{\infty}$ and $\{S_n^{(k)}(x)\}_{n=0}^{\infty}$ be the sequences, respectively, of Fibonacci, Fibonacci-type and polynacci polynomials of order k. Then

$$f_{n+1}^{(k)}(x) = \sum_{\substack{n_1, \ldots, n_k \\ n_1 + 2n_2 + \ldots + kn_k = n}} \binom{n_1 + \ldots + n_k}{n_1, \ldots, n_k} x^{k(n_1+\ldots+n_k)-n}, \quad n \geq 0; \qquad (2.6)$$

$$F_{n+1}^{(k)}(x) = \sum_{\substack{n_1, \ldots, n_k \\ n_1 + 2n_2 + \ldots + kn_k = n}} \binom{n_1 + \ldots + n_k}{n_1, \ldots, n_k} x^{n_1+\ldots+n_k}, \quad n \geq 0; \qquad (2.7)$$

$$S_{n+1}^{(k)}(x) = \sum_{\substack{n_1, \ldots, n_k \\ n_1 + 2n_2 + \ldots + kn_k = n}} \binom{n_1 + \ldots + n_k}{n_1, \ldots, n_k} p^n (q/p)^{n_1+\ldots+n_k}, \quad n \geq 0. \qquad (2.8)$$

Remark 2.2: Relations (2.6) and (2.7), respectively, were first derived in [6] and [7], whereas (2.8) is new.

Remark 2.3: Setting $x = 1$ in (2.6), we get a multinomial expansion of $f_{n+1}^{(k)}$ ($n \geq 0$), where $\{f_n^{(k)}\}_{n+0}^{\infty}$ denotes the sequence of Fibonacci numbers of order k. This expansion was first obtained in [5], and it was recently used by Putz [8] as "the link" between the Fibonacci numbers of order k and his Pascal polytopes.

3. POLYNACCI POLYNOMIALS AND CONSECUTIVE SUCCESSES

In this section we relate the polynacci polynomials of order k to the number of Bernoulli trials N_k until the occurrence of the k-th consecutive success, and as a corollary we reestablish a recurrence of [1] and [4]. We also give new proofs of some results of [3], [6] and [7], and, in a sense, unify them.

Proposition 3.1: Let $\{S_n^{(k)}(p)\}_{n=0}^{\infty}$ be the sequence of polynacci polynomilas of order k, and denote by N_k the number of Bernoulli trials until the first occurrence of the k-th consecutive success.

Set

$$b_n^{(k)}(p) = P(N_k=n), \quad n \geq 0. \tag{3.1}$$

Then

$$b_{n+k}^{(k)}(p) = p^k S_{n+1}^{(k)}(p), \quad n \geq 0. \tag{3.2}$$

Proof: It follows directly, by comparison of (2.8) and Theorem 3.1 of Philippou and Muwafi [5].

We proceed, however, to give a second proof, which is of some interest in its own right, and part of which will also be needed for showing Corollary 3.2 below. Let $E_{k,n}$ be the event that no k consecutive successes occur in the first n Bernoulli trials, and set $P_k(n) = P(E_{k,n})$, $n \geq 0$. Then

$$P_k(n) = \begin{cases} 1, & 0 \leq n \leq k-1, \\ P\left[\bigcup_{i=1}^{k} E_{k,n-i} \cap \{\underbrace{fss \ldots s}_{i-1}\} \right] = \sum_{i=1}^{k} P_k(n - i)qp^{i-1}, & n \geq k, \end{cases} \tag{3.3}$$

where "f" denotes failure and "s" denotes success. It is also easily seen that

$$P(N_k = n + k + 1) = P(E_{k,n} \cap \{\underbrace{fss \ldots s}_{k}\}) = qp^k P_k(n), \quad n \geq 0. \tag{3.4}$$

In order to complete the proof of the proposition, it suffices to show that

$$qP_k(n) = S_{n+2}^{(k)}(p), \quad n \geq 0.$$

This can be easily done by mathematical induction, using relation (3.3).

To Proposition 3.1 there is the following corollary, which can be easily obtained by means of (3.2).

<u>Corollary 3.1</u>: Let $\{b_n^{(k)}(p)\}_{n+0}^{\infty}$ be as in (3.1). Then

$$b_n^{(k)}(p) = q \sum_{i=1}^{k} p^{i-1} b_{n-i}^{(k)}(p), \quad n \geq k. \tag{3.5}$$

We also have the following.

<u>Corollary 3.2</u>: Let $\{b_n^{(k)}(p)\}_{n=0}^{\infty}$ be as in (3.1). Then

$$b_n^{(k)}(p) = \begin{cases} 0, & 0 \leq n \leq k-1, \\ p^k, & n = k, \\ qp^k \left[1 - \sum_{i=k}^{n-k-1} b_i^{(k)}(p) \right], & n \geq k+1. \end{cases} \tag{3.6}$$

<u>Proof</u>: Trivially, $b_n^{(k)}(p) = 0$ $(0 \leq n \leq k - 1)$ and $b_k^{(k)}(p) = p^k$ by (3.1). Also, $b_n^{(k)}(p) = qp^k$ for $k + 1 \leq n \leq 2k$, by means of (3.1) and (3.4), which shows (3.6) for $k+1 \leq n \leq 2k$. Next we observe that

$$b_n^{(k)}(p) = b_{n-1}^{(k)}(p) - qp^k b_{n-1-k}^{(k)}(p), \quad n \geq 2k + 1,$$

by means of (3.5), from which the proof of (3.6) is easily completed.

<u>Remark 3.1</u>: Relation (3.6) was derived from first principles in [1]. Another version of it was independently found in [4], by means of the Fibonacci-type polynomials of order k.

Next we obtain the following result, by means of (2.4) and (3.2).

Corollary 3.3: Let $\{b_n^{(k)}(p)\}_{n=0}^{\infty}$ be as in (3.1), and denote by $g_4(s, p)$ its generating function. Then $g_4(s, p)$ is analytic for $|s| \leq 1$, and

$$g_4(s, p) = \frac{p^k s^k}{1-qs(1 + ps + \ldots + p^{k-1}s^{k-1})} . \tag{3.7}$$

Remark 2.1 and Corollary 3.3 imply

Corollary 3.4: Let $g_4(s, p)$ be as in Corollary 3.3, and let $g_i(s, x)$ $(1 \leq i \leq 3)$ be as in Remark 2.1. Then $g_1(s, x)$, $g_2(s, x)$ and $g_3(s, x)$ are analytic in $|s| \leq 1$, $|s| \leq p$ and $|s| \leq p(q/p)^{1/k}$, respectively, and

$$g_4(s, p) = p^k s^{k-1} g_3(s, p) \tag{3.8}$$

$$= p^{k-1} s^{k-1} g_2(ps, q/p)$$

$$= p^{k-1}(p/q)^{1/k} s^{k-1} g_1(ps(q/p)^{1/k}, (q/p)^{1/k}),$$

for $|s| < \min \{p, p(q/p)^{1/k}\}$.

Obviously, Corollary 3.4 relates the sequences $\{f_n^{(k)}(x)\}_{n=0}^{\infty}$, $\{F_n^{(k)}(x)\}_{n=0}^{\infty}$ and $\{S_n^{(k)}(x)\}_{n=0}^{\infty}$. Furthermore, equating the coefficients of s^{n+k} in (3.8), and using also (3.2), the following corollary is derived.

Corollary 3.5: Let $\{b_n^{(k)}(p)\}_{n=0}$ be as in (3.1), and let $S_n^{(k)}(p)$, $f_n^{(k)}(x)$ and $F_n^{(k)}(x)$, respectively, be the polynacci, Fibonacci and Fibonacci-type polynomials of order k. Then

$$b_{n+k}^{(k)}(p) = p^k S_{n+1}^{(k)}(p) \tag{3.9a}$$

$$= p^{n+k}(q/p)^{n/k} f_{n+1}^{(k)}\left[(q/p)^{1/k}\right] \tag{3.9b}$$

$$= p^{n+k} F_{n+1}^{(k)} (q/p) \tag{3.9c}$$

<u>Remark</u> 3.2:   Relation (3.9a) is of course relation (3.2). It was first obtained in [3],
             while relations (3.9b) and (3.9c) were first derived in [6] and [7],
             respectively.

REFERENCES

[1] Aki, S., Kuboki, H., & Hirano, K. "On Discrete Distributions of Order k." *Annal*
     of the Institute of Statistical Mathematics 36, No. 3 (1984): pp 431-440.

[2] Levesque, C. "On m-th Order Linear Recurrences." *The Fibonacci Quarterly*
     23, No. 4 (1985): pp 290-295.

[3] Philippou, G. N. "Fibonacci Polynomials of Order k and Probability
     Distributions of order k." *Ph. D. Thesis (in Greek). University of Patras,*
     Patras, Greece (1984).

[4] Philippou, A. N., & Makri, F. S. "Longest Success Runs and Fibonacci-type
     Polynomials." *The Fibonacci Quarterly 23, No. 4* (1985): pp 338-345.

[5] Philippou, A. N. & Muwafi, A. A. "Waiting for the kth Consecutive Success an
     the Fibonacci Sequence of Order k." *The Fibonacci Quarterly 20, No. 1* (1982):
     pp 28-32.

[6] Philippou, A. N., Georghiou, C., & Philippou, G. N. "Fibonacci Polynomials of
     Order k, Multinomial Expansions, and Probability." *International Journal of*
     Mathematics and Mathematical Sciences 6, No. 3 (1983): pp 545-550.

[7] Philippou, A. N., Georghiou, C., & Philippou, G. N. "Fibonacci-Type Polynomials
     of Order k with Probability Applications." *The Fibonacci Quarterly 23, No. 2*
     (1985): pp 100-105.

[8] Putz, J. F. "The Pascal Polytope: An Extension of Pascal's Triangle to N
     Dimensions." *The College Mathematics Journal 17, No. 2* (1986): pp 144-155.

[9] Shane, H. D. "A Fibonacci Probability Function." *The Fibonacci Quarterly 11,*
     No. 6 (1973): pp 511-522.

[10] Stanley, R. P. "Generating Functions." *In Studies in Combinatorics. Gian-Carlo*
     Rota Edr., MAA Studies in Mathematics, Vol. 17 (1978): pp 100-141.

Herta T. Freitag and Piero Filipponi

ON THE REPRESENTATION OF INTEGRAL SEQUENCES {Fₙ/d} AND {Lₙ/d} AS SUMS OF FIBONACCI NUMBERS AND AS SUMS OF LUCAS NUMBERS

1. SCOPE AND PURPOSE OF THE STUDY

Based on Zeckendorf's theorem concerning the unique sum-representation of any positive integer in terms of Fibonacci numbers as well as Lucas numbers /1/, the purpose of this study is the development of relationships which enable prediction of the NUMBER of addends in these representations. Integral sequences $\{F_n/d\}$ and $\{L_n/d\}$ are considered such that d, with $2 \leq d \leq 20$, is a predetermined integer and n is subject to appropriate conditions to assure integral elements in these sequences. Restrictions on n such that $F_n \equiv 0 \pmod{d}$ can always be determined. However, for n ε {5, 8, 10, 12, 13, 15, 16, 17, 20} there does not exist an n-value such that $L_n \equiv 0 \pmod{d}$.

Symbolizing the number of F-addends in the sum representation of N by f(N) and the corresponding one for L-addends by $\ell(N)$, functions $f(F_n/d)$, $\ell(F_n/d)$, $f(L_n/d)$ and $\ell(L_n/d)$ are developed. Finally, a few generalizations are introduced.

Conjectures which exhibit a recursive behavior become apparent through a study of early cases of n. Validations of these conjectures constitute the proofs. In some cases, series which - by invoking the Binet form - become geometric series are involved. The method of proof will be shown by typical examples. Results only will be given for the other theorems. A list of basic relationships used in the proofs is appended.

2. Theorems

2.1 Simplest results. Linear Functions in k with integral coefficients, (d ε {4, 19}).

Theorem 1

(i) If $n \equiv 0 \pmod{18}$, i.e., n = 18k, k a positive integer, then
$f(F_n/19) = \ell(F_n/19) = n/6$, i.e., $f(F_{18k}/19) = \ell(F_{18k}/19) = 3k$.

97

N. Philippou et al. (eds.), Applications of Fibonacci Numbers, 97–112.
1988 by Kluwer Academic Publishers.

(ii) If $n \equiv 9 \pmod{18}$, i.e., $n = 18k+9$, k a nonnegative integer, then
$f(L_n/19) = 2n/9$, ie., $f(L_{18k+9}/19) = 4k+2$ and
$\ell(L_n/19) = (n-3)/6$, i.e., $\ell(L_{18k+9})/19 = 3k+1$.

Proof:

(i) Conjecture 1 (see Chart 1)

$$F_{18k}/19 = S_1 + F_{18(k-1)}/19 \text{ where } S_1 = \sum_{i=1}^{3} F_{18k-13+2i} - F_{18k-6} - F_{18k-12}.$$

Let $18k-9 = t$. Then: $F_{t+9} - F_{t-9} \overset{?}{=} 19(F_{t+3} - F_{t-3})$.

By relationship (7), the conjecture holds. As this conjecture displays a recursive pattern, $f(F_{18k}/19) = 3k$ may be immediately seen by mathematical induction.

Chart 1. (for Theorem 1)

k $F_{18k}/19$	F-represent.	$f(F_{18k}/19)$	L-representation	$\ell(F_{18k}/19)$
1 136	$\sum_{i=1}^{3} F_{5+2i}$	$1 \cdot 3$	$L_{10}+L_5+L_0$	$3 = 1 \cdot 3$
2 785808	$(\sum_{i=1}^{3} F_{23+2i})+$ $+F_{18}/19$	$2 \cdot 3$	$L_{28}+L_{23}+L_{19}+$ $+L_{15}+ L_{12}+L_8$	$6 = 2 \cdot 3$
3 4540398488	$(\sum_{i=1}^{3} F_{41+2i})+$ $+F_{36}/19$	$3 \cdot 3$	$L_{46}+(\sum_{i=1}^{3} L_{29+4i})+$ $+L_{30}+L_{26}+F_{18}/19$	$6+3=3 \cdot 3$

Conjecture 2 (see Chart 1). For $k \geq 3$,

$$F_{18k}/19 = L_{18k-8}+S_2+L_{18k-24}+L_{18k-28}+F_{18(k-2)}/19, \text{ where, by (15), } S_2 =$$

$$\sum_{i=1}^{3} L_{18k-25+4i} - \{(L_{18k-9}-L_{18k-13})-(L_{18k-21}-L_{18k-25})\}/5. \text{ Let } 18k-18 = t. \text{ Then:}$$

$$F_{t+18}-F_{t-18} \overset{?}{=} 19\{L_{t+10}+L_{t-10})+(L_{t+9}-L_{t+5})/5-(L_{t-3}-L_{t-7})/5+L_{t-6}\}, \text{ or}$$

by (9) and (10), $136L_t \overset{?}{=} (L_{t+10}+L_{t-10})+(F_{t+7}+F_{t-7})$.

Now, by invoking relationships (6) and (3), we obtain

$(L_{t+10}+L_{t-10})+(F_{t+7}+F_{t-7})=L_t(L_{10}+F_7) = 136L_t$.

The conjecture holds and the recursive nature of the proven conjecture leads to $\ell(F_{18k}/19) = 3k$.

(ii) Part 1:

Conjecture 3 (see Chart 2). For $k \geq 1$,

$L_{18k+9}/19 = S_3+L_{18(k-1)+9}/19$, where $S_3 = \sum_{i=1}^{4}F_{18k-8+3i} = A+B$ with $A = F_{18k+1}+F_{18k-5}$ and $B = F_{18k+4}+F_{18k-2}$. But, by (3), $A+B = 2(L_{18k-2}+L_{18k+1})$ $=4L_{18k}$. Now, by (8), $L_{18k+9}-L_{18k-9} = 76L_{18k}$ and thus the conjecture holds and $f(L_{18k+9}/19 = 4k+2$.

Chart 2. (for Theorem 1)

k	$L_{18k+9}/19$	F-represent.	$f(L_{18k+9}/19)$	L-represent.	$\ell(L_{18k+9}/19)$
0	4	F_4+F_2	$2=0\cdot4+2$	L_3	$1=0\cdot3+1$
1	23116	$(\sum_{i=1}^{4}F_{10+3i})+$ $+L_9/19$	$4+2=1\cdot4+2$	$(\sum_{i=1}^{3}L_{14+2i})+$ $+L_9/19$	$3+1=1\cdot3+1$
2	133564244	$(\sum_{i=1}^{4}F_{28+3i})+$ $+L_{27}/19$	$2.4+2$	$(\sum_{i=1}^{3}L_{32+2i})+$ $+L_{27}/19$	$2\cdot3+1$

Part 2:

Conjecture 4 (see Chart 2). For $k \geq 1$,

$L_{18k+9}/19 = S_4+L_{18(k-1)+9}/19$, where $S_4 = \sum_{i=1}^{3}L_{18k-4+2i}=L_{18k+3}-L_{18k-3}$.

Hence, by (8), $S_4 - 4L_{18k}$.

As, again by (8), $L_{18k+9}-L_{18k-9} = 76L_{18k}$, the conjecture is verified and,

by Chart 2 and the recursive nature of the established conjecture,
$\ell(L_{18k+9}/19) = 3k+1$.

Theorem 2

(i) If $n \equiv 0(\mod 6)$, i.e., $n = 6k$, k a positive integer, then
$f(F_n/4) = \ell(F_n/4) = n/6$, i.e., $f(F_{6k}/4) = \ell(F_{6k}/4) = k$.

(ii) If $n \equiv 3(\mod 6)$, i.e., $n = 6k+3$, k a nonnegative integer, then
$f(L_n/4) = n/3$, i.e., $f(L_{6k+3}/4) = 2k+1$, and $\ell(L_n/4) = (n+3)/6$, i.e.,
$\ell(L_{6k+3}/4) = k+1$.

2.2 Results dependent on parity of k (i.e., modulus 2 is the determining factor), ($d \; \varepsilon \; \{6, 9, 14, 18, 20\}$).

Theorem 3

(i) If $n \equiv 0(\mod 12)$, i.e., $n = 12k$, k a positive integer, then

$$f(F_n/18) = \begin{cases} (n-8)/4, & \text{if } n \equiv 12(\mod 24) \\ n/4, & \text{if } n \equiv 0(\mod 24) \end{cases}, \text{i.e., } f(F_{12k}/18) = \begin{cases} 3k-2, & (k \text{ odd}) \\ 3k, & (k \text{ even}), \end{cases}$$

$$\ell(F_n/18) = \begin{cases} (n+4)/8, & \text{if } n \equiv 12(\mod 24) \\ n/8, & \text{if } n \equiv 0(\mod 24) \end{cases}, \text{i.e., } \ell(F_{12k}/18) = [(3k+1)/2],$$

where $[\cdot]$ denotes the greatest integer function.

(ii) If $n \equiv 6(\mod 12)$, i.e., $n = 12k+6$, k a nonnegative integer, then

$$f(L_n/18) = \begin{cases} (7n+18)/24, & \text{if } n \equiv 18(\mod 24) \\ (7n-18)/24, & \text{if } n \equiv 6(\mod 24) \end{cases}, \text{i.e., } f(L_{12k+6}/18) = \begin{cases} (7k+5)/2, & (k \text{ odd}) \\ (7k+2)/2, & (k \text{ even}) \end{cases}.$$

$$\ell(L_n/18) = \begin{cases} (n+6)/4, & \text{if } n \equiv 18(\mod 24) \\ (n-2)/4, & \text{if } n \equiv 6(\mod 24) \end{cases}, \text{i.e., } \ell(L_{12k+6}/18) = \begin{cases} 3k+3, & (k \text{ odd}) \\ 3k+1, & (k \text{ even}) \end{cases}.$$

Proof:

(i) Part 1:

Conjecture 5 (see Chart 3). For $k \geq 2$,

$$F_{12k}/18 = S_1 + F_{12(k-2)}/18, \text{ where } S_1 = \sum_{t=1}^{6} F_{12k-19+2t} = F_{12k-6} - F_{12k-18}.$$

Let $12k-12 = t$. Then $F_{t+12} - F_{t-12} \overset{?}{=} 18(F_{t+6} - F_{t-6})$.

By relationship (9), both sides are seen to be equal to $144L_t$. Thus, the conjecture holds and we may, without distinguishing the parity cases, state that $f(F_{12k}/18) = 6[k/2] + ((-1^{k+1}+1)/2$. The theorem follows.

Chart 3 (for Theorem 3)

k	$F_{12k}/18$	F-representation	$f(F_{12k}/18)$	L-represent.	$\ell(F_{12k}/18)$
1	8	F_6	$1=0 \cdot 6+1$	$L_4 + L_1$	$2=0 \cdot 3+2$
2	2576	$\sum_{t=1}^{6} F_{5+2t}$	$6=1 \cdot 6$	$\sum_{t=1}^{3} L_{4+4t}$	$3=1 \cdot 3$
3	829464	$(\sum_{t=1}^{6} F_{17+2t}) + F_{12}/18$	$6+1=1 \cdot 6+1$	$(\sum_{t=1}^{3} L_{16+4t}) + F_{12}/18$	$3+2=1 \cdot 3+2$

Part 2:
Conjecture 6 (see Chart 3). For $k \geq 2$,

$$F_{12k}/18 = S_2 + F_{12(k-2)}/18, \text{ where } S_2 = \sum_{t=1}^{3} L_{12k-20+4t}. \text{ By (15)}$$
$$S_2 - \{(L_{12k-4} - L_{12k-8}) - (L_{12k-16} - L_{12k-20})\}/5 - F_{12k-6} - F_{12k-18}.$$

Hence, analogous to Part 1, the conjecture holds. Thus,

$$\ell(F_{12k}/18) = [k/2] \cdot 3 + \begin{cases} 2, \text{ if } k \text{ is odd} \\ 0, \text{ if } k \text{ is even.} \end{cases}$$

The theorem follows.
(ii) is proved similarly.

The results for theorems involving $d \in \{6, 9, 14\}$ are shown in Chart 4.

Chart 4

Theorem 4

d = 6	n = 12k	$f(F_n/6)$	$\ell(F_n/6)$
		(7k-3)/2, if k is odd 7k/2 , if k is even	$[(5k+1)/2]$
		$f(L_n/6)$	$\ell(L_n/6)$
	n = 12k+6	(7k+5)/2, if k is odd (7k+2)/2, if k is even	$[(7k+3)/2]$

Theorem 5

d = 9	n = 12k	$f(F_n/9)$	$\ell(F_n/9)$
		3k-1, if k is odd 3k, if k is even	2k+1, if k is odd 2k, if k is even
		$f(L_n/9)$	$\ell(L_n/9)$
	n = 12k+6	$[(7k+2)/2]$	3k+3, if k is odd 3k+1, if k is even

Theorem 6

d = 14	n = 24k	$f(F_n/14)$	$\ell(F_n/14)$
		7k-2, if k is odd 7k, if k is even	6k+1, if k is odd 6k, if k is even
		$f(L_n/14)$	$\ell(L_n/14)$
	n = 24k+12	$[(13k+5)/2]$	7k+5, if k is odd 7k+3, if k is even

Chart 4 (continued)

Theorem 7

d = 20	n = 30k	f(F_n/20)	ℓ(F_n/20)
		[15k/2]	6k+1, if k is odd 6k, if k is even
	$\exists$ n such that $L_n \equiv$ 0(mod 20)		

2.3 Results dependent on modulus 4 (d ε {5, 10, 13, 17}).

Theorem 8

If $n \equiv$ 0(mod 9), i.e., n = 9k, k a positive integer, then

$$f(F_n/17) = \begin{cases} 5n/18, & \text{if } n \equiv 0(\text{mod } 36) \\ (5n-27)/18, & \text{if } n \equiv 9(\text{mod } 36) \\ (5n-54)18, & \text{if } n \equiv 18(\text{mod } 36) \text{ , i.e.,} \\ (5n-9)/18, & \text{if } n \equiv 27(\text{mod } 36) \end{cases}$$

$$f(F_{9k}/17) = \begin{cases} 5k/2, & \text{if } k \equiv 0(\text{mod } 4) \\ (5k-3)/2, & \text{if } k \equiv 1(\text{mod } 4) \\ (5k-6)/2, & \text{if } k \equiv 2(\text{mod } 4) \\ (5k-1)/2, & \text{if } k \equiv 3(\text{mod } 4), \quad \text{and} \end{cases}$$

$$\ell(F_n/17) = \begin{cases} n/4, & \text{if } n \equiv 0(\text{mod } 36) \\ (n-5)/4, & \text{if } n \equiv 9(\text{mod } 36) \\ (n-10)/4, & \text{if } n \equiv 18(\text{mod } 36), \text{ i.e.,} \\ (n+9)/4, & \text{if } n \equiv 27(\text{mod } 36) \end{cases}$$

$$\ell(F_{9k}/17) = \begin{cases} 9k/4, & \text{if } k \equiv 0(\text{mod } 4) \\ (9k-5)/4, & \text{if } k \equiv 1(\text{mod } 4) \\ (9k-10)/4, & \text{if } k \equiv 2(\text{mod } 4) \\ (9k+9)/4, & \text{if } k \equiv 3(\text{mod } 4). \end{cases}$$

Proof:

Part 1:

Conjecture 7 (see Chart 5). For $k \geq 5$,

$F_{9k}/17 = F_{9k-6}+S_1+S_2+F_{9k-31}+F_{9(k-4)}/17$, where, by (11)

$$S_1 = \sum_{i=1}^{5}F_{9k-23+2i} = F_{9k-12}-F_{9k-22} \text{ and}$$

$$S_2 = \sum_{i=1}^{3}F_{9k-30+2i} = F_{9k-23}-F_{9k-29}.$$

Let $9k-18 = t$. Then

$$F_{t+18}-F_{t-18} \overset{?}{=} 17\{F_{t+12}-F_{t-12})+(F_{t+6}-F_{t-6})\}.$$

Then, by relationship (9), both sides are equivalent to $17 \cdot 152L_t$. Thus, the conjecture holds and

$$f(F_{9k}/17) = [(k+1)/4] \cdot 10 + \begin{cases} 0, & \text{if } k \equiv 0 \pmod 4 \\ 1, & \text{if } k \equiv 1 \pmod 4 \\ 2, & \text{if } k \equiv 2 \pmod 4 \\ -3, & \text{if } k \equiv 3 \pmod 4. \end{cases}$$

The theorem follows.

Part 2:

Conjecture 8 (see Chart 5). For $k \geq 5$,

$F_{9k}/17 = L_{9k-8}+S_3+L_{9k-28}+F_{9(k-4)}/17$, where by (15)

$$S_3 = \sum_{i=1}^{7}L_{9k-26+2i} = L_{9k-11}-L_{9k-25}. \text{ Let } 9k-18 = t. \text{ Then}$$

$$F_{t+18}-F_{t-18} \overset{?}{=} 17\{(L_{t+10}+L_{t-10})+(L_{t+7}-L_{t-7})\}.$$

Using relationships (6) and (8), both sides are seen to equal $17 \cdot 152L_t$. Thus, the conjecture holds and

$$\ell(F_{9k}/17) = [(k+1)/4]\cdot 9 + \begin{cases} 0, & \text{if } k \equiv 0 \text{ or } 3\pmod 4 \\ 1, & \text{if } k \equiv 1\pmod 4 \\ 2, & \text{if } k \equiv 2\pmod 4. \end{cases}$$

But this is only another form of our theorem.

$\not\exists$ n such that $L_n \equiv 0\pmod{17}$.

Chart 5. (for Theorem 8)

k	$F_{gk}/17$	F-representation	$f(F_{gk}/17)$	L-represent.	$\ell(F_{gk}/17)$
1	2	F_3	$1=0\cdot10+1$	L_0	$1=0\cdot9+1$
2	152	$F_{12}+F_6$	$2=0\cdot10+2$	$L_{10}+L_7$	$2=0\cdot9+2$
3	11554	$F_{21}+\sum_{i=1}^{6}F_{2+2i}$	$7=0\cdot10+7$	$L_{19}+\sum_{i=1}^{8}L_{2i-1}$	$9=1\cdot9+0$
4	878256	$F_{30}+\left[\sum_{i=1}^{5}F_{13+2i}\right]+$ $+F_{12}+F_{10}+F_8+F_5$	$10=1\cdot10+0$	$L_{28}+\left[\sum_{i=1}^{7}L_{2i+10}\right]$ $+L_8$	$9=1\cdot9+0$
5	66759010	$F_{39}+\left[\sum_{i=1}^{5}F_{22+2i}\right]+F_{21}+$ $+F_{19}+F_{17}+F_{14}+F_9/17$	$10+1=1\cdot10+1$	$L_{37}+\left[\sum_{i=1}^{7}L_{2i+19}\right]+$ $+L_{17}+F_9/17$	$9+1=1\cdot9+1$

The other d-values leading to results which involve modulus 4 throughout are $d \varepsilon \{5, 10, 13\}$. Chart 6 displays these theorems. $\not\exists$ n-values such that $L_n \equiv 0\pmod{\overline{d}}$ where $\overline{d} \varepsilon\{5, 10, 13\}$.

Chart 6

Theorem 9 (d=5, n=5k)	Theorem 10 (d=10, n=15k)	Theorem 11 (d=13, n=7k)
$f(F_n/5)$	$f(F_n/10)$	$f(F_n/13)$
$3k/2$, if $k \equiv 0 \pmod 4$	$4k$, if $k \equiv 0 \pmod 4$	$2k$, if $k \equiv 0 \pmod 4$
$(3k-1)/2$, if $k \equiv 1 \pmod 4$	$4k-1$, if $k \equiv \begin{cases} 1 \pmod 4 \\ 2 \pmod 4 \end{cases}$	$2k-1$, if $k \equiv 1 \pmod 4$
$(3k-2)/2$, if $k \equiv 2 \pmod 4$		$2k-2$, if $k \equiv 2 \pmod 4$
$(3k+1)/2$, if $k \equiv 3 \pmod 4$	$4k+1$, if $k \equiv 3 \pmod 4$	$2k+1$, if $k \equiv 3 \pmod 4$
$\ell(F_n/5)$	$\ell(F_n/10)$	$\ell(F_n/13)$
$5k/4$, if $k \equiv 0 \pmod 4$	$9k/2$, if $k \equiv 0 \pmod 4$	$7k/4$, if $k \equiv 0 \pmod 4$
$(5k-1)/4$, if $k \equiv 1 \pmod 4$	$(9k-3)/2$, if $k \equiv 1 \pmod 4$	$(7k-3)/4$, if $k \equiv 1 \pmod 4$
$(5k-6)/4$, if $k \equiv 2 \pmod 4$	$(9k-6)/2$, if $k \equiv 2 \pmod 4$	$(7k-10)/4$, if $k \equiv 2 \pmod 4$
$(5k+5)/4$, if $k \equiv 3 \pmod 4$	$(9k+1)/2$, if $k \equiv 3 \pmod 4$	$(7k+7)/4$, if $k \equiv 3 \pmod 4$

2.4. Miscellaneous (d ε { 2, 3, 7, 8, 11, 12, 15, 16}.

Chart 7 displays the results of theorems involving the remaining d-values i.e., d ε {2, 3, 7, 8, 11, 12, 15, 16}. These show mixed patterns where more than one of the characteristic features displayed above occur for the same divisor d.

$\not\exists$ n such that $L_n \equiv 0 \pmod{\bar{d}}$ where $\bar{d}$ ε {8, 12, 15, 16}.

Chart 7a   (d ε {2, 3, 7, 11})

Theorem 12

n	$f(F_n/2)$	$\ell(F_n/2)$	$f(L_n/2)$	$\ell(L_n/2)$
3k	k	$3k/4$, if $k \equiv 0 \pmod 4$ $(3k+1)/4$, if $k \equiv 1 \pmod 4$ $(3k-2)/4$, if $k \equiv 2 \pmod 4$ $(3k+3)/4$, if $k \equiv 3 \pmod 4$	k	k

Theorem 13

n	$f(F_n/3)$	$\ell(F_n/3)$	n	$f(L_n/3)$	$\ell(L_n/3)$
4k	k	$[(k+1)/2]$	4k+2	k+1	k+1

Theorem 14

n	$f(F_n/7)$	$\ell(F_n/7)$	n	$f(L_n/7)$	$\ell(L_n/7)$
8k	2k-1, if k is odd 2k, if k is even	k	8k+4	$[(5k+3)/2]$	2k+2, if k is odd 2k+1, if k is even

Theorem 15						
d	n	$f(F_n/11)$	$\ell(F_n/11)$	n	$f(L_n/11)$	$\ell(L_n/11)$
11	10k	k	$[(3k+1)/2]$	10k+5	2k+1	k+1

Chart 7b (d ε {8, 12, 15, 16})

Theorem 16			
d	n	$f(F_n/8)$	$\ell(F_n/8)$
8	6k	k	$[(k+1)/2]$

Theorem 17			
d	n	$f(F_n/12)$	$\ell(F_n/12)$
12	12k	$[(5k+1)/2]$	2k

Theorem 18			
d	n	$f(F_n/15)$	$\ell(F_n/15)$
15	20k	5k	$[(7k+1)/2]$

Theorem 19			
d	n	$f(F_n/16)$	$\ell(F_n/16)$
16	12k	$[5k/2]$	2k

2.5. A glimpse at some generalizations

A few general relationships were also obtained. In most cases conjecture exhibiting a recursive behavior again formed the bases of these proofs. In on case, an algebraic relationship pertaining to binomials could be used.

Theorem 20: If n is an even positive integer and k any positive integer, then

$$f(F_{kn}/F_n) = \begin{cases} 1, & \text{if } n = 2 \\ k, & \text{if } n \geq 4 \end{cases} \quad \text{and} \quad \ell(F_{kn}/F_n) = [(k+1)/2].$$

Proof: Part 1: $f(F_{kn}/F_n)$

    (i) if $n = 2$, the result is trivial.

    (ii) $n \geq 4$:

Conjecture: $k \geq 2$

$$F_{kn}/F_n = L_{(k-1)n} + F_{(k-2)n}/F_n.$$

By letting $(k-1)n = t$ and applying relationship (9), this conjecture is seen to be correct.

Case 1: k is odd. Here,

$$F_{kn}/F_n = (\sum_{i=1}^{(k-1)/2} L_{2in}) + 1 = \sum_{i=1}^{(k-1)/2} (F_{2in+1} + F_{2in-1}) + 1. \text{ Thus,}$$

$$f(F_{kn}/F_n) = 1 + 2(k-1)/2 = k.$$

Case 2: k is even. Now,

$$F_{kn}/F_n = \sum_{i=1}^{k/2} L_{(2i-1)n} = \sum_{i=1}^{k/2} (F_{(2i-1)n+1} + F_{(2i-1)n-1}). \text{ Thus,}$$

$$f(F_{kn}/F_n) = 2k/2 = k.$$

Proof: Part 2: $\ell(F_{kn}/F_n)$

    (i) $n = 2$

Case 1: k is odd

$$F_{2k} = L_1 + \sum_{i=1}^{(k-1)/2} L_{4i}, \text{ hence } \ell(F_{2k}) = (k+1)/2.$$

Case 2: k is even

$$F_{2k} = \sum_{i=1}^{k/2} L_{4i-2}, \text{ resulting in } \ell(F_{2k}) = k/2.$$

Thus, for all k-values, $\ell(F_{2k}) = \left[(k+1)/2\right]$.

(ii) $n \geq 4$. From the above, $\ell(F_{kn}/F_n) = (k-1)/2+1 = (k+1)/2$ for odd k's and - for even k-values - it becomes $k/2$. Thus, for all k-values, $\ell(F_{kn}/F_n) = [(k+1)/2]$.

Theorem 21: If n is an odd positive integer and k an even one, then

$$f(F_{kn}/L_n) = \begin{cases} 1, & \text{if } n = 1 \\ k/2, & \text{if } n \geq 3 \end{cases} \quad \text{and}$$

$$\ell(F_{kn}/L_n) = \begin{cases} [(k+3)/4], & \text{if } n = 1 \\ [(n+3)/4], & \text{for } k = 2 \\ (n+1)/2, & \text{for } k = 4 \\ \left.\begin{array}{l} (3n+5)/4, & \text{if } n \equiv 1 \pmod 4 \\ (3n+3)/4, & \text{if } n \equiv 3 \pmod 4 \end{array}\right\} \text{ for } k = 6 \end{cases} \left.\right\}, \text{ if } n \geq 3 .$$

Thus far, a further generalization for even k-values, $k \geq 8$, has not been completed.

Theorem 22: If n and k are both odd positive integers, then

$$f(L_{kn}/L_n) = \begin{cases} 2, & \text{if } n = 1 \\ k, & \text{if } n \geq 3 \end{cases} \quad \text{and} \quad \ell(L_{kn}/L_n) = \begin{cases} 1, & \text{if } n = 1 \\ (k+1)/2, & \text{if } n \geq 3 \end{cases} .$$

3. CONCLUDING REMARKS

While the authors have undertaken this study primarily in response to their interest in mathematical relationships, it has been pointed out /6/ that these functions - by virtue of their property of being one-way functions-may be of considerable interest to cryptanalysts. The authors hope to continue this study as the results obtained have - as could be predicted - opened a host of further avenues for investigations.

4. RELATIONSHIPS USED IN THE PROOFS OF THE THEOREMS

$$F_{t+m} = F_m F_{t-1} + F_{m+1} F_t \tag{1}$$

$$L_{t+m} = F_m L_{t-1} + F_{m+1} L_t \tag{2}$$

$$F_{t+m} + F_{t-m} = L_t F_m, \text{ m odd} \tag{3}$$

$$L_{t+m} + L_{t-m} = 5 F_t F_m, \text{ m odd} \tag{4}$$

$$F_{t+m} + F_{t-m} = F_t L_m, \text{ m even} \tag{5}$$

$$L_{t+m} + L_{t-m} = L_t L_m, \text{ m even} \tag{6}$$

$$F_{t+m} - F_{t-m} = F_t L_m, \text{ m odd} \tag{7}$$

$$L_{t+m} - L_{t-m} = L_t L_m, \text{ m odd} \tag{8}$$

$$F_{t+m} - F_{t-m} = L_t F_m, \text{ m even} \tag{9}$$

$$L_{t+m} - L_{t-m} = 5 F_t F_m, \text{ m even} \tag{10}$$

$$\sum_{i=1}^{r} F_{ai+b} = \frac{F_{a(r+1)+b} + (-1)^{a-1} F_{ar+b} - (F_{a+b} + (-1)^{a-1} F_b)}{L_a + (-1)^{a-1} - 1} \tag{11}$$

Special cases: $a = 1 \rightarrow F_{b+2+r} - F_{b+2}$ (12)

$a = 2 \rightarrow F_{b+1+2r} - F_{b+1}$ (13)

$a = 4 \rightarrow (L_{b+2+4r} - L_{b+2})/5$ (14)

$$\sum_{i=1}^{r} L_{ai+b} = \frac{L_{a(r+1)+b} + (-1)^{a-1} L_{ar+b} - (L_{a+b} + (-1)^{a-1} L_b)}{L_a + (-1)^{a-1} - 1} \tag{15}$$

Special cases: $a = 1 \rightarrow L_{b+2+r} - L_{b+2}$ (16)

$a = 2 \rightarrow L_{b+1+2r} - L_{b+1}$ (17)

$a = 4 \rightarrow F_{b+2+4r} - F_{b+2}$. (18)

REFERENCES

[1] J.L. Brown, Jr. "Zeckendorf's Theorem and Some Applications." *The Fibonacci Quarterly 2, No. 3* (1964): pp 163-168.

[2] D.A. Klarner. "Partitions of N into Distinct Fibonacci Numbers." *The Fibonacci Quarterly 6, No. 4* (1968): pp 235-243.

[3] V.E. Hoggatt, Jr. *Fibonacci and Lucas Numbers*. Houghton Mifflin Co., Boston (1969)

[4] P. Filipponi. "Sulle Proprieta dei Rapporti fra Particolari Numeri di Fibonacci e di Lucas." *Note Recensioni Notizie 33, No. 3-4* (1984): pp 91-96.

[5] P. Filipponi. "The Representation of Certain Integers as a Sum of Distinct Fibonacci Numbers." *Fondazione Ugo Bordoni, Int. Report 2B0985*, Roma (1985).

[6] Herta T. Freitag, and P. Filipponi. "On the Representation of Integers in Terms of Sums of Lucas Numbers." *Note Recensioni Notizie 34, No. 3* (1985): pp 145-150.

Lawrence Somer

PRIMES HAVING AN INCOMPLETE SYSTEM OF RESIDUES FOR A CLASS OF SECOND-ORDER RECURRENCES

1. INTRODUCTION

Shah [4] and Bruckner [1] showed that if p is a prime and p > 7, then the Fibonacci sequence {F_n} has an incomplete system of residues modulo p. Shah established this result for the cases in which $p \equiv 1, 9, 11$, or 19 modulo 20, while Bruckner proved the result true for the remaining cases in which $p \equiv 3$ or 7 modulo 10. Burr [2] extended these results by determining all the positive integers m for which the Fibonacci sequence has an incomplete system of residues modulo m.

We will generalize these results by considering all linear recurrences of the form

$$w_{n+2} = aw_{n+1} + bw_n , \tag{1}$$

denoted by w(a, b) or (w), where the initial terms w_0 and w_1 are integers, a is an integer, and b = ±1. Let $D = a^2 + 4b$ be the discriminant of w(a, b). We say w(a, b) is defective modulo m if w(a, b) has an incomplete system of residues modulo m. We will try to determine all those primes p for which all recurrences w(a, b) are defective modulo p, where b is a fixed integer equal to 1 or -1, and a varies over all integers such that $D = a^2 + 4b \not\equiv 0 \pmod{p}$. Our results are given in Theorems 1 and 2.

Theorem 1: Let p be a prime.

    (1) If $p \geq 5$, then all recurrences w(a, -1) for which $D = a^2 - 4 \not\equiv 0$ modulo p are defective modulo p.

113

.N. Philippou et al. (eds.), Applications of Fibonacci Numbers, 113–141.
1988 by Kluwer Academic Publishers.

(2) For each of $p = 2$ or 3, there exists a recurrence w (a, -1) with D $\not\equiv 0$ (mod p) such that (w) is non-defective modulo p. In fact, w(3, -1) with initial terms $w_0 = 0$ and $w_1 = 1$ and discriminant 5 is non-defective modulo the primes 2 and 3.

(3) If w(a, -1) is a recurrence such that $D \equiv 0$ (mod p) and $w_0 \equiv 0$, $w_1 \equiv 1$ (mod p), then $a \equiv \pm 2$ (mod p) and w(a, -1) has a complete system of residues modulo p. In particular,

$$w_n \equiv n \text{ or } w_n \equiv (-1)^{n+1} n \text{ (mod p)}.$$

Theorem 2: Let p be a prime.

(1) If $p > 7$ and $p \not\equiv 1$ or 9 (mod 20), then all recurrences w(a, 1) for which $D = a^2 + 4 \not\equiv 0$ (mod p) are defective modulo p.

(2) For each of $p = 2, 3, 5,$ or 7, there exists a recurrence w(a, 1) with discriminant $D \not\equiv 0$ (mod p) such that (w) is non-defective modulo p. In fact, the Fibonacci sequence with discriminant 5 is non-defective modulo the primes 2, 3, and 7. The Pell sequence w(2, 1) with initial terms 0 and 1 and discriminant 8 is non-defective modulo 5.

(3) If w(a, 1) is a recurrence such that $D \equiv 0$ (mod p) and $w_0 \equiv 0$, $w_1 \equiv 1$ (mod p), then $a \equiv \pm 2i$ (mod p), where $i^2 \equiv -1$ (mod p) and w(a, 1) has a complete system of residues modulo p. In particular,

$$w_n \equiv ni^{n-1} \text{ or } w_n \equiv n(-i)^{n-1} \text{ (mod p)}.$$

2. PRELIMINARIES

Suppose the recurrence w(a, ± 1) is non-defective modulo p. Then $w_n \equiv$ (mod p) for some n. Suppose further that $w_{n+1} \equiv 0$ (mod p). It is well-known that w(a, ± 1) is purely periodic modulo p (see [8], page 312). Hence, $w_n \equiv 0$ (mod p) for all n, and (w) cannot have a complete system of residues modulo p, contrary to assumption. Thus, $w_n \equiv 0$ (mod p) and $w_{n+1} \equiv r$ (mod p) for some non-negative integer n and some non-zero residue r modulo p. If $\{w_n\}$ is non-defective modulo p, then, clearly, so is the recurrence $\{sw_n\}$, where $s \not\equiv 0$ (mod p). Now consider the non-defective recurrence $\{ r^{-1}w_n \}$ modulo p. Then

$$r^{-1}w_n \equiv 0, \; r^{-1}w_{n+1} \equiv 1 \; (\text{mod p}).$$

Since $\{r^{-1}w_n\}$ is purely periodic modulo p, then any translate of $\{r^{-1}w_n\}$ is non-defective modulo p. Thus, without loss of generality, we can assume that if $w(a, \pm1)$ is non-defective modulo p, then $w_0 \equiv 0$, $w_1 \equiv 1$ (mod p). Let $u(a, b)$ be the recurrence satisfying the recursion relation (1) with initial terms $u_0 = 0$, $u_1 = 1$. The recurrence (u) is called a Lucas sequence of the first kind (LSFK). Thus, to prove that for a prime p and fixed integer $b = \pm1$ all recurrences $w(a, b)$ with discriminant $D \not\equiv 0$ (mod p) are defective modulo p, we need only show that all recurrences $u(a, b)$ with $D \not\equiv 0$ (mod p) are defective modulo p. Hence, to prove Theorems 1 and 2, we need to consider only LSFK's $u(a, \pm1)$.

The following definitions and known results concerning LSFK's $u(a, b)$ will be necessary for our proofs of Theorems 1 and 2. Unless stated otherwise, b is an integer not necessarily equal to ±1.

Definition 1: Let $u(a, b)$ be a LSFK. Let p be a prime such that $p \nmid b$. Then the period of $u(a, b)$ modulo p, denoted by $\mu(a, b, p)$, is the least positive integer t such that

$$u_{n+t} \equiv u_n \; (\text{mod p})$$

for all non-negative integers n.

Definition 2: Let $u(a, b)$ be a LSFK. Let p be a prime such that $p \nmid b$. Then the restricted period of $u(a, b)$ modulo p or rank of apparition of p in $u(a, b)$, denoted by $\alpha(a, b, p)$ is the least positive integer t such that

$$u_{n+t} \equiv su_n \; (\text{mod p})$$

for all non-negative integers n and some non-zero residue s. Then s = s(a, b, p) is called the <u>multiplier</u> of u(a, b) modulo p. Let $\beta(a, b, p)$ denote the <u>exponent of the multiplier</u> modulo p.

<u>Remark</u>: It is clear that $\alpha(a, b, p)$ is the least positive integer t such that $u_t \equiv 0 \pmod{p}$. It is also easy to see that

$$\alpha(a, b, p) \mid \mu(a, b, p)$$

and

$$\beta(a, b, p) = \mu(a, b, p)/\alpha(a, b, p).$$

Further, it clearly follows that

$$u_{k\alpha(a,b,p) + n} \equiv s^k u_n \pmod{p}$$

for a fixed non-negative integer k and all non-negative integers n.

<u>Proposition 1</u>: Let u(a, b) be a LSFK. Then

$$u_{m+n} = bu_m u_{n-1} + u_n u_{m+1}$$

<u>Proof</u>: This is easily proved by induction. □

<u>Proposition 2</u>: Let b be a non-zero integer and let u(a, b) be a LSFK which is also defined for negative subscripts. Then

$$u_{-n} = (-1)^{n+1} (u_n/b^n)$$

for $n \geq 0$.

<u>Proof</u>: This is easily proved by induction. □

<u>Proposition 3</u>: Let u(a, b) be a LSFK with discriminant D. Let p be an odd prime such that $p \nmid b$. Then

$$u_{p-(D/p)} \equiv 0 \ (\text{mod } p) \text{ and } \alpha(a, b, p) \mid p\text{-}(D/p),$$

where (D/p) denotes the Legendre symbol. Moreover,

$$u_p \equiv (D/p) \ (\text{mod } p).$$

If (-b/p) = 1 and $p \nmid D$, then

$$\alpha(a, b, p) \mid \tfrac{1}{2}(p\text{-}(D/p)).$$

If (-b/p) = -1 and $p \nmid D$, then

$$\alpha(a, b, p) \nmid \tfrac{1}{2}(p\text{-}(D/p)).$$

Proof: This is proved in [3]. □

Proposition 4: Let u(a, b) be a LSFK. Let p be an odd prime such that $p \nmid b$. If (D/p) = 1,

$$\mu(a, b, p) \mid p\text{-}1.$$

Proof: This is proved in [8], page 313. □

Proposition 5: Let u(a, b) be a LSFK. Let p be an odd prime such that $p \nmid b$. Let h denote the exponent of -b modulo p. Let H denote the least common multiple of $\alpha(a, b, p)$ and h. Then

$$\mu(a, b, p) = H \text{ or } 2H.$$

Proof: This is proved by Wyler [10]. □

Proposition 6: Let u(a, -1) be a LSFK. Let p be an odd prime. Then

$$\beta(a, -1, p) = 1 \text{ or } 2.$$

If $(D/p) = -1$,

$$\mu(a, -1, p) \mid p+1$$

Proof: By Proposition 5, it follows that

$$\mu(a, -1, p) = \alpha(a, -1, p) \text{ or } 2\alpha(a, -1, p),$$

and hence,

$$\beta(a, -1, p) = \mu(a, -1, p)/ \alpha(a, -1, p) = 1 \text{ or } 2.$$

By Proposition 3,

$$\alpha(a, -1, p) \mid \tfrac{1}{2}(p+1).$$

Since $\mu(a, -1, p) = \beta(a, -1, p) \, \alpha(a, -1, p)$, it now follows that

$$\mu(a, -1, p) \mid p+1. \quad \square$$

Proposition 7: Let u(a, 1) be a LSFK. Let p be an odd prime such that $p \nmid b$. Then

$$\beta(a, 1, p) = 1, 2, \text{ or } 4.$$

If $(D/p) = -1$ and $p \equiv 3 \pmod 4$,

$$\beta(a, 1, p) = 2, 4 \mid \alpha(a, 1, p), \text{ and } \alpha(a, 1, p) \mid p+1, \text{ but } \alpha(a, 1, p) \nmid$$
$$\tfrac{1}{2}(p+1).$$

If $(D/p) = -1$ and $p \equiv 1 \pmod 4$,

$$\beta(a, 1, p) = 4, \ \alpha(a, 1, p) \equiv 1 \pmod 2, \text{ and } \alpha(a, 1, p) \mid \tfrac{1}{2}(p+1).$$

Proof: By Proposition 5, it follows that

$$\mu(a, 1, p) = \alpha(a, 1, p), 2\alpha(a, 1, p), \text{ or } 4\alpha(a, 1, p),$$

and thus,

$$\beta(a, 1, p) = 1, 2, \text{ or } 4.$$

If $p \equiv 3 \pmod 4$ then $(-1/p) = -1$, and it follows from Proposition 3 that $\alpha(a, 1, p) \mid p+1$, but $\alpha(a, 1, p) \nmid \frac{1}{2}(p+1)$. If $p \equiv 1 \pmod 4$, then $(-1/p) = 1$, and it follows from Proposition 3 that $\alpha(a, 1, p) \mid \frac{1}{2}(p+1)$. The rest of the theorem is proved in [7], pages 325-326. □

Proposition 8: Let $v(a, b)$ be a Lucas sequence of the second kind (LSSK) defined by the recursion relation

$$v_{n+2} = av_{n+1} + bv_n \,,$$

$v_0 = 2$, $v_1 = a$. If r is a fixed positive integer, then the derived sequence $\{u_{nr}/u_r\}$ is a LSFK $u(a^{(r)}, b^{(r)})$, where $a^{(r)} = v_r$ and $-b^{(r)} = (-b)^r$.

Proof: This is proved by Lehmer [3]. □

Proposition 9: Let $p \equiv 1 \pmod 4$ be a prime. Then the number of distinct quadratic residues c^2 such that $c^2 + 4$ is congruent to a quadratic non-residue is $(p - 1)/4$.

Proof: In [6], page 39, it was shown that the number of distinct quadratic residues d^2 such that $d^2 + 4$ is congruent to a quadratic residue is $(p + 3)/4$. Since there are $(p + 1)/2$ quadratic residues including 0, it follows there are

$$(p + 1)/2 - (p + 3)/4 = (p - 1)/4$$

distinct quadratic residues c^2 such that $c^2 + 4$ is congruent to a quadratic non-residue. □

The following lemmas which are new results will also be needed for the proofs of Theorems 1 and 2.

Lemma 1: Consider the LSFK $u(a, b)$. Let p be a prime such that $p \nmid b$. Then the ratios u_{n+1}/u_n are all distinct modulo p for $1 \leq n \leq \alpha(a, b, p) - 1$.

Proof: By the remark after Definition 2, $u_n \not\equiv 0 \pmod{p}$ for $1 \leq n \leq \alpha(a, b, p) - 1$. Thus, all the ratios u_{n+1}/u_n are defined for $1 \leq n \leq \alpha(a, b, p) - 1$. Assume

$$u_{c+1}/u_c \equiv u_{d+1}/u_d \pmod{p}$$

for $1 \leq c \leq d \leq \alpha(a, b, p) - 1$. We will show $c = d$. Then

$$u_{c+1}u_d \equiv u_{d+1}u_c \pmod{p}. \tag{2}$$

By (2) and the recursion relation defining (u),

$$(au_c + bu_{c-1})\, u_d \equiv (au_d + bu_{d-1})\, u_c \pmod{p}$$

This implies

$$bu_c u_{d-1} \equiv bu_d u_{c-1} \pmod{p}.$$

Since $b \not\equiv 0 \pmod{p}$, this implies that

$$u_c u_{d-1} \equiv u_d u_{c-1} \pmod{p}.$$

Continuing in this manner for $c - 1$ steps, we obtain

$$u_1 u_{d-c} \equiv u_{d-c+1} u_0 \pmod{p}.$$

Since $u_0 \equiv 0$, $u_1 \equiv 1$, and $u_{d-c+1} \not\equiv 0 \pmod{p}$, it follows that $u_{d-c} \equiv 0 \pmod{p}$. Thus, $d - c = 0$ and $d = c$. ◻

Lemma 2: Consider the LSFK $u(a, b)$. Let p be a prime such that $p \nmid b$. Let r be a fixed integer such that $1 \leq r \leq \alpha(a, b, p) - 1$. Then the ratios u_{n+r}/u_n

are all distinct modulo p for $1 \leq n \leq \alpha(a, b, p) -1$.

Proof: Since $\alpha(a, b, p)$ is the first positive integer t such that $u_t \equiv 0 \pmod{p}$, all the ratios u_{n+r}/u_n are well-defined modulo p. Assume

$$u_{c+r}/u_c \equiv u_{d+r}/u_d \pmod{p} \tag{3}$$

for $1 \leq c \leq d \leq \alpha(a, b, p) -1$. We will show c = d. From (3) we have

$$u_{c+r}u_d \equiv u_{d+r}u_c \pmod{p} \tag{4}$$

From Proposition 1,

$$u_{c+r} = bu_r u_{c-1} + u_c u_{r+1}, \quad u_{d+r} = bu_r u_{d-1} + u_d u_{r+1} \tag{5}$$

From (4) and (5) we obtain

$$(bu_r u_{c-1} + u_c u_{r+1})u_d \equiv (bu_r u_{d-1} + u_d u_{r+1})u_c \pmod{p}$$

or

$$bu_r u_{c-1}u_d \equiv bu_r u_{d-1}u_c \pmod{p}.$$

Since $b \not\equiv 0$ and $u_r \not\equiv 0 \pmod{p}$, this implies that

$$u_c u_{d-1} \equiv u_d u_{c-1} \pmod{p} \tag{6}$$

By the proof of Lemma 1, (6) implies that c = d. □

Remark: The statements and proofs of Lemmas 1 and 2 are immediate generalizations of results obtained by Bruckner [1], pages 217-219.

Lemma 3: Let u(a, b) be a LSFK. Let p be an odd prime such that $p \nmid b$. Suppose that $1 \leq n \leq \alpha(a, b, p) -2$. Let $t = \alpha(a, b, p)$. Let s be the multiplier of u(a, b) modulo p. Then

$$(u_{n+1}/u_n)\,(u_{t-n}/u_{t-n-1}) \equiv -b \pmod{p}$$

<u>Proof</u>: We proceed by induction. We claim the lemma is true for $n = 1$. Note that $u_1 = 1$ and $u_2 = a$. Also, $u_t \equiv 0$, $u_{t+1} \equiv s \pmod{p}$. Thus, by the recursion relation defining $u(a, b)$,

$$bu_{t-1} + au_t \equiv bu_{t-1} + a \cdot 0 \equiv s \pmod{p}$$

or

$$u_{t-1} \equiv s/b \pmod{p}.$$

Also,

$$u_{t-2} \equiv (u_t - au_{t-1})/b \equiv (0 - asb^{-1})/b \equiv -as/b^2 \pmod{p}.$$

Hence,

$$(u_2/u_1)\,(u_{t-1}/u_{t-2}) \equiv (a/1)\,(sb^{-1}/(-asb^{-2})) \equiv -b \pmod{p}.$$

Now suppose the hypothesis is true up to $n = k$. Let

$$u_{k+1}/u_k \equiv r \pmod{p}.$$

Then

$$u_{t-k}/u_{t-k-1} \equiv -b/r \pmod{p}.$$

Thus,

$$u_{k+1} \equiv ru_k \text{ and } u_{t-k} \equiv (-b/r)u_{t-k-1} \pmod{p}.$$

Hence, by the recursion relation defining $u(a, b)$,

$$u_{k+2}/u_{k+1} \equiv (au_{k+1} + bu_k)/u_{k+1} \equiv (ar + b)u_k\,/(ru_k) \equiv (ar + b)/r \pmod{p}.$$

Also,

$$u_{t-k-1}/u_{t-k-2} \equiv u_{t-k-1}/(b^{-1}(u_{t-k} - au_{t-k-1}))$$

$$\equiv u_{t-k-1}\,/(b^{-1}(-br^{-1} - a)\,u_{t-k-1}) \equiv rb/(-b - ar) \pmod{p}.$$

Hence,

$$(u_{k+2}/u_{k+1})\,(u_{t-k-1}/u_{t-k}) \equiv ((ar + b)/r)\,(rb/(-b - ar)) \equiv -b \pmod{p},$$

and the proof is complete by induction. □

Remark: Lemma 3 was given in [5], page 323 for the case of the Fibonacci sequence.

Lemma 4: Let u(a, b) be a LSFK. Let p be an odd prime such that $p \nmid b$. Let r be a
fixed positive integer such that $1 \leq r \leq \alpha(a, b, p) - 2$. Let n be a positive
integer such that $n + r \leq \alpha(a, b, p) - 1$. Let $t = \alpha(a, b, p)$. Then

$$(u_{n+r}/u_n)(u_{t-n}/u_{t-n-r}) \equiv (-b)^r \pmod{p}.$$

Proof: Let
$$R_n \equiv u_{n+1}/u_n \pmod{p}$$

for $1 \leq n \leq t - 2$. By Lemma 3 it follows that

$$R_n R_{t-n-1} \equiv -b \pmod{p}.$$
Hence,
$$(u_{n+r}/u_n)(u_{t-n}/u_{t-n-r}) \equiv [R_n R_{n+1} \ldots R_{n+r-1}][R_{t-n-1}R_{t-n-2} \ldots R_{t-n-r}]$$

$$\equiv (R_n R_{t-n-1})(R_{n+1}R_{t-n-2}) \ldots (R_{n+r-1}R_{t-n-r}) \equiv (-b)^r \pmod{p}. \ \square$$

Lemma 5: Let u(a, -1) be a LSFK and p be a prime. Let $t = \alpha(a, -1, p)$.

(i) If $\beta(a, -1, p) = 1$, then

$$u_{t-k} \equiv -u_k \pmod{p}$$

for $1 \leq k \leq t - 1$.

(ii) If $\beta(a, -1, p) = 2$, then

$$u_{t-k} \equiv u_k \pmod{p}$$

for $1 \leq k \leq t - 1$.

Proof: (i) If $\beta(a, -1, p) = 1$, then

$$s(a, -1, p) \equiv 1 \equiv u_{t+1} \pmod{p}.$$

Hence, by the recursion relation defining $u(a, -1)$,

$$(-1)u_{t-1} \equiv u_{t+1} - au_t \equiv 1 - a \cdot 0 \equiv 1 \pmod{p}$$
or
$$u_{t-1} \equiv -1 \equiv -u_1 \pmod{p}.$$
Also,
$$u_{t-2} \equiv (-1)(u_t - au_{t-1}) \equiv (-1)(0 - a(-1)) \equiv -a \equiv -u_2 \pmod{p}.$$

The assertion now follows by induction.

(ii) If $\beta(a, -1, p) = 2$, then

$$s(a, -1, p) \equiv -1 \equiv u_{t+1} \pmod{p}.$$
Thus,
$$u_{t-1} \equiv (-1)(u_{t+1} - au_{t-1}) \equiv (-1)(-1 - a \cdot 0) \equiv 1 \equiv u_1 \pmod{p}$$
and
$$u_{t-2} \equiv (-1)(u_t - au_{t-1}) \equiv (-1)(0 - a \cdot 1) \equiv a \equiv u_2 \pmod{p}$$

The assertion now follows by induction. □

Lemma 6: Let $u(a, 1)$ be a LSFK and p be a prime. Let $t = \alpha(a, 1, p)$.

(i) If $\beta(a, 1, p) = 1$, then

$$u_{t-k} \equiv (-1)^{k+1}u_k \pmod{p}$$

for $1 \leq k \leq t - 1$.

(ii) If $\beta(a, 1, p) = 2$, then

$$u_{t-k} \equiv (-1)^k u_k \pmod{p}$$

for $1 \leq k \leq t - 1$.

(iii) If $\beta(a, 1, p) = 4$, then

$$u_{2t-k} \equiv (-1)^k u_k \pmod{p}$$

for $1 \leq k \leq 2t - 1$.

Proof: (i) If $\beta(a, 1, p) = 1$, then

$$s(a, 1, p) \equiv 1 \equiv u_{t+1} \pmod{p}.$$

Hence,

$$u_{t-1} \equiv u_{t+1} - au_t \equiv 1 - a \cdot 0 \equiv 1 \equiv u_1 \pmod{p}$$

and

$$u_{t-2} \equiv u_t - au_{t-1} \equiv 0 - a \cdot 1 \equiv -a \equiv -u_2 \pmod{p}$$

Assertion (i) now follows by induction.

(ii) If $\beta(a, 1, p) = 2$, then

$$s(a, 1, p) \equiv -1 \equiv u_{t+1} \pmod{p}.$$

Thus,

$$u_{t-1} \equiv u_{t+1} - au_t \equiv -1 - a \cdot 0 \equiv -1 \equiv -u_1 \pmod{p}$$

and

$$u_{t-2} \equiv u_t - au_{t-1} \equiv 0 - a(-1) \equiv a \equiv u_2 \pmod{p}$$

The assertion now follows by induction.

(iii) If $\beta(a, 1, p) = 4$, let $s(a, 1, p) \equiv s \pmod{p}$. Then $s^4 \equiv 1 \pmod{p}$ and $s^2 \equiv -1 \pmod{p}$. By the remark following Definition 2,

$$u_{2t+n} \equiv (-1)u_n \pmod{p}$$

for all $n \geq 0$. Hence,

$$u_{2t-1} \equiv u_{2t+1} - au_{2t} \equiv s^2 \cdot 1 - a \cdot 0 \equiv -1 - 0 \equiv -1 \equiv -u_1 \pmod{p}$$

and

$$u_{2t-2} \equiv u_{2t} - au_{2t-1} \equiv 0 - a(-1) \equiv a \equiv u_2 \pmod{p}.$$

The assertion now follows the induction. $\square$

Lemma 7: Let $u(a, 1)$ be a LSFK and p be an odd prime. Let $t = \alpha(a, 1, p)$.

(i) If $(D/p) = -1$ and $p \equiv 3 \pmod{4}$, then $t \equiv 0 \pmod{4}$ and

$$u_n \not\equiv \pm u_{n+2r} \pmod{p}$$

for any positive integers n and r such that either $n + 2r \leq t/2$ or it is the case that $n \geq t/2$ and $n + 2r \leq t - 1$.

(ii) If $(D/p) = -1$ and $p \equiv 1 \pmod{4}$, then $t \equiv 1 \pmod{2}$ and

$$u_n \not\equiv \pm u_{n+2r} \pmod{p}$$

for any positive integers n and r such that $n + 2r \leq t - 1$.

Proof: (i) Suppose $(D/p) = -1$ and $p \equiv 3 \pmod{4}$. Then, by Proposition 7, $4 \mid t$. Suppose there exist positive integers n and r such that $n + 2r \leq t - 1$ and

$$u_n \equiv \pm u_{n+2r} \pmod{p}.$$

Then

$$u_{n+2r}/u_n \equiv \pm 1 \pmod{p}.$$

By Lemma 4,

$$(u_{n+2r}/u_n)(u_{t-n}/u_{t-n-2r}) \equiv (-1)^{2r} \equiv 1 \pmod{p}$$

and

$$u_{t-n}/u_{t-n-2r} \equiv u_{n+2r}/u_n \equiv \pm 1 \pmod{p}.$$

Thus, by Lemma 2,

$$n + 2r = t - n$$

or

$$n = (t/2) - r.$$

Hence,

$$n = (t/2) - r \text{ and } n + 2r = (t/2) + r.$$

The assertion now follows.

(ii) Suppose $(D/p) = 1$ and $p \equiv 1 \pmod 4$. Then by Proposition 7, $t \equiv 1 \pmod 2$. Suppose there exist positive integers n and r such that $n + 2r \leq t - 1$ and

$$u_n \equiv \pm u_{n+2r} \pmod{p}.$$

Then

$$u_{n+2r}/u_n \equiv \pm 1 \pmod{p}.$$

Again, By Lemma 4,

$$(u_{n+2r}/u_n)(u_{t-n}/u_{t-n-2r}) \equiv (-1)^{2r} \equiv 1 \pmod{p}$$

and

$$u_{t-n}/u_{t-n-2r} \equiv u_{n+2r}/u_n \equiv \pm 1 \pmod{p}.$$

Hence, by Lemma 2,

$$n + 2r = t - n$$

and

$$n = (t - 2r)/2.$$

However, this is impossible since $t \equiv 1 \pmod 2$, and the result follows. $\square$

Lemma 8: Let u(a, 1) be a LSFK. Let $p \equiv 3 \pmod 4$ be a prime such that $(D/p) = -1$. Let $t = \alpha(a, 1, p)$. Suppose there exist positive integers n and r such that $n + 2r - 1 \le t/2$ and

$$u_{n+2r-1}/u_n \equiv \pm 1 \pmod p.$$

Then the only positive integers m such that $m \le t - 1$ and $u_m \equiv \pm u_n$ are

$$m = n, \; m = n + 2r - 1, \; m = t - n - 2r + 1, \text{ and } m = t - n.$$

Proof: This follows from Lemmas 7, 4, and 2. □

Lemma 9: Let u(a, 1) be a LSFK. Let $p \equiv 3 \pmod 4$ be a prime such that $(D/p) = -1$. Let $t = \alpha(a, 1, p)$. Then $t \equiv 0 \pmod 4$. Suppose there exist positive integers n and r such that $n < t$, $2r - 1 < t$, and

$$u_{n+2r-1}/u_n \equiv \pm 1 \pmod p. \tag{7}$$

(i) Suppose in (7), $n < t/2$ and $t/2 < n + 2r - 1 < t - 1$. Suppose

$$u_c \equiv \pm u_d \equiv \pm u_n \pmod p,$$

where $1 \le c < d < 3t/2$ and $d - c = 2m - 1 < t$ for some positive integer m. If $1 \le c < d < t/2$ or $t/2 \le c < d < t - 1$, then

$$2m - 1 = |\, t - 2n - 2r + 1 \,| < 2r - 1.$$

If $1 \le c < t/2 < d \le (t/2) - 1$, then

$$2m - 1 = 2r - 1.$$

If $t/2 \le c < t < d < t + 2r - 1$, then

$$2m - 1 = t - (2r - 1) \,.$$

(ii) Suppose in (7), $t/2 \leq n < t < n + 2r - 1 < t + 2r - 1$. Suppose

$$u_c \equiv \pm u_d \equiv \pm u_n \pmod{p},$$

where $1 \leq c < d < t + 2r - 1$ and $d - c = 2m - 1 < t$ for some positive integer m. If $1 \leq c < d \leq t/2$ or $t/2 \leq c < d \leq t - 1$, then

$$2m - 1 = \mid 2t - 2n - 2r + 1 \mid < 2r - 1.$$

If $1 \leq c < t/2 < d \leq (t/2) - 1$, then

$$2m - 1 = t - (2r - 1).$$

If $t/2 \leq c < t < d \leq t + 2r$, then

$$2m - 1 = 2r - 1.$$

Proof: By Proposition 7, $t \equiv 0 \pmod 4$ and $\beta(a, 1, p) = 2$. Then $s(a, 1, p) \equiv -1 \pmod p$ and $u_{t+k} \equiv -u_k \pmod p$ for all $k \geq 0$.

(i) By the above argument and Lemma 8, it follows that the only subscripts i such that $1 \leq i < 3t/2$ and

$$u_i \equiv \pm u_n \pmod{p}$$

are $i = n, t - n - 2r - 1, n + 2r - 1, t - n, t + n$, and $2t - n - 2r - 1$.

The result now follows by inspection.

(ii) By Lemma 8 and using the fact that $s(a, 1, p) \equiv -1 \pmod p$, it follows that the only subscripts i such that $1 \leq i \leq t + 2r$ and

$$u_i \equiv \pm u_n \pmod{p}$$

are $i = n + 2r - 1 - t$, $t - n$, n, $2t - n - 2r + 1$, $n + 2r - 1$, and $2t - n$.

The rest of the assertion follows by inspection. □

3. PROOFS OF THE MAIN ASSERTIONS

We are now ready for the proofs of Theorems 1 and 2.

<u>Proof</u> <u>of</u> <u>Theorem</u> <u>1</u>: (1) Let $N(a, -1, p)$ denote the number of distinct residues of $u(a, -1)$ modulo p. If $(D/p) = 1$, then by Proposition 4, $\mu(a, -1, p) \mid p-1$. Hence, $N(a, -1, p) \leq p - 1$, and $u(a, -1)$ is defective modulo p in this case.

Now suppose $(D/p) = -1$. Then by Proposition 3, $\alpha(a, -1, p) \mid \frac{1}{2}(p + 1)$. By Proposition 6, $\beta(a, -1, p) = 1$ or 2. Hence,

$$\mu(a, -1, p) \mid p + 1.$$

Thus, in order to have $N(a, -1, p) = p$, we must have that $\mu(a, -1, p) = p+1$ which can only occur if $\alpha(a, -1, p) = (p + 1)/2$ and $\beta(a, -1, p) = 2$. Thus, $s(a, -1, p) \equiv -1 \pmod{p}$ and

$$u_{((p+1)/2)+n} \equiv -u_n \pmod{p}$$

for $1 \leq n \leq (p + 1)/2$. Also, by Lemma 5

$$u_{((p+1)/2)-k} \equiv u_k \pmod{p}$$

for $1 \leq k \leq (p + 1)/4$. Moreover, $u_0 \equiv 0 \pmod{p}$. Therefore,

$$N(a, -1, p) \leq (2(p + 1)/4) + 1 = (p + 3)/2 < p$$

if $p > 3$. Assertion (1) now follows.

(2) This follows by inspection.

(3) Suppose $D = a^2 - 4 \equiv 0 \pmod{p}$. Then $a^2 \equiv 4$ and $a \equiv \pm 2 \pmod{p}$.

The rest follows by induction. $\square$

<u>Proof</u> of <u>Theorem</u> 2: (1) Let $N(a, 1, p)$ be the number of distinct residues of $u(a, 1)$ modulo p. First suppose $(D/p) = 1$. Then by Proposition 4, $\mu(a, 1, p) \mid p-1$. Thus, $N(a, 1, p) < p$, and $u(a, 1)$ is defective modulo p.

Now suppose $(D/p) = -1$. We first consider the case $p \equiv 3 \pmod{4}$. Then by Proposition 7,

$$\alpha(a, 1, p) \mid p+1 \text{ but } \alpha(a, 1, p) \nmid \tfrac{1}{2}(p+1), \text{ and } \beta(a, 1, p) = 2.$$

Thus, the only possibility for $\mu(a, 1, p) \geq p$ occurs if $\alpha(a, 1, p) = p + 1$. In this case, $\mu(a, 1, p) = 2(p + 1)$ and $s(a, 1, p) \equiv -1 \pmod{p}$. Then

$$u_{p+1+k} \equiv (-1)u_k \pmod{p} \tag{8}$$

for $1 \leq k \leq (p + 1)$. Moreover, by Lemma 6 (ii),

$$u_{p+1-k} \equiv (-1)^k u_k \pmod{p} \tag{9}$$

for $1 \leq k \leq (p + 1)/2$. Let N_1 be the largest t such that there exist integers $n_1, n_2, \ldots, n_t$ for which $1 \leq n_i \leq (p + 1)/2$ and $u_{n_i} \not\equiv \pm u_{n_j}$ if $1 \leq i < j \leq (p + 1)/2$. Since $u_0 \equiv 0 \pmod{p}$, it follows from (8) and (9) that

$$N(a, 1, p) = 2N_1 + 1.$$

Thus, we can show $u(a, 1)$ to be defective modulo p if we can show that $N_1 \leq ((p + 1)/2) - 2$. We can show that $N_1 \leq (p + 1)/2) - 2$ if we can find integers c, d and e, f such that

$$u_c \equiv \pm u_d \text{ and } u_e \equiv \pm u_f \pmod{p},$$

where $1 \leq c < d \leq (p + 1)/2$, $1 \leq e < f \leq (p + 1)/2$, and either $c \neq e$ or $d \neq f$. We note that by Lemma 7, we must have that both d - c and f - e are odd integers.

Consider the residues

$$u_{n+1}/u_n \pmod{p}$$

for $1 \leq n \leq p$. By Lemma 1, these residues are distinct and thus consist of all the residues modulo p. Hence, there exists a positive integer $i \leq p$ such that

$$u_{i+1}/u_i \equiv 1 \pmod{p}.$$

By Lemma 3, it follows that there exists a positive integer $m \leq ((p + 1)/2)$ -1 such that

$$u_{m+1}/u_m \equiv \pm 1 \pmod{p}. \tag{10}$$

Hence, $u_{m+1} \equiv \pm u_m \pmod{p}$, and we have that $N_1 \leq ((p + 1)/2) - 1$.

Now consider the sets of residues $\{u_{n+3}/u_n\}$ and $\{u_{n+5}/u_n\}$, where $1 \leq n \leq p$. By Lemma 1, each of these sets consists of all the p distinct residues modulo p. Thus, there exist positive integers $c \leq p$ and $d \leq p$ such that

$$u_{c+3}/u_c \equiv 1 \pmod{p} \tag{11}$$

and

$$u_{d+5}/u_d \equiv 1 \pmod{p}. \tag{12}$$

If we have that either $c + 3 \leq (p + 1)/2$ or $d + 5 \leq (p + 1)/2$, then observe that

$$u_{c+3} \equiv u_c \pmod{p}$$

or

$$u_{d+5} \equiv u_d \ (\text{mod } p).$$

This implies that $N_1 \leq ((p + 1)/2) - 2$ and thus, $u(a, 1)$ is defective modulo p.

Now assume that either

$$(p + 1)/2 \leq c < c + 3 < p + 1 \qquad\qquad (13)$$

or

$$(p + 1)/2 \leq d < d + 5 < p + 1. \qquad\qquad (14)$$

Let $c_1 = (p + 1) - c$ and $d_1 = (p + 1) - d$. If (13) or (14) hold, then

$$4 \leq c_1 = (p + 1) - c \leq (p + 1)/2$$

or

$$6 \leq d_1 = (p + 1) - d \leq (p + 1)/2.$$

By Lemma 4, we then have that

$$u_{c_1}/u_{c_1-3} \equiv -1 \ (\text{mod } p)$$

or

$$u_{d_1}/u_{d_1-5} \equiv -1 \ (\text{mod } p)$$

This again implies that $N_1 \leq ((p + 1)/2) - 2$, and $u(a, 1)$ is defective modulo p.

Now assume the LSFK $u(a, 1)$ is non-defective modulo p for $p > 7$.
Thus, in congruence (11), we must have that either

$$1 \leq c < (p + 1)/2 < c + 3 \leq p + 1 \qquad\qquad (15)$$

or

$$(p + 1)/2 < c < p + 1 < c + 3 \leq p + 3. \qquad\qquad (16)$$

Similarly, in (12), it must follow that

$$1 \leq d < (p + 1)/2 < d + 5 < p + 1 \qquad\qquad (17)$$

or

$$(p + 1)/2 < d < p + 1 < d + 5 \leq p + 5. \tag{18}$$

If (11) holds and either (15) or (16) holds, then the proof of Lemma 9 implies that there exists an integer n_1 such that $1 \leq n_1 \leq ((p + 1)/2) - 1$ and

$$u_{n_1} \equiv \pm u_{n_1+1} \equiv \pm u_c \pmod{p}. \tag{19}$$

If (12) holds and either (17) or (18) holds, then the proof of Lemma 9 implies that there exist integers n_2 and r such that

$$1 \leq n_2 < n_2 + 2r - 1 \leq (p + 1)/2$$

and

$$u_{n_2} \equiv \pm u_{n_2+2r-1} \equiv \pm u_d \pmod{p}. \tag{20}$$

Moreover, by the proof of Lemma 9, we must have that $2r - 1 = 1$ or $2r - 1 = 3$. Since $u(a, 1)$ is non-defective modulo p, it follows that in congruences (11) and (12),

$$u_c \equiv \pm u_m \text{ and } u_d \equiv \pm u_m \pmod{p}, \tag{21}$$

where m is defined as in congruence (10). It then follows by Lemma 9 that it is either the case that inequalities (15) and (18) both hold or it is the case that inequalities (16) and (17) both hold. But then congruence (21) and Lemma 9 together imply that $5 + 3 = p + 1$. However, this is impossible since $p > 7$. This contradiction establishes that $u(a, 1)$ is defective modulo p if $(D/p) = -1$, $p \equiv 3 \pmod 4$, and $p > 7$.

Finally, we consider the case in which $(D/p) = -1$, $p \equiv 1 \pmod 4$, $p \not\equiv 1$ or $9 \pmod{20}$, and $p > 7$. Note that by the law of quadratic reciprocity, $(5/p) = -1$. Further, by Proposition 7,

$$\alpha(a, 1, p) \equiv 1 \pmod 2, \ \alpha(a, 1, p) \mid (p + 1)/2, \text{ and } \beta(a, 1, p) = 4.$$

Let $s \equiv s(a, 1, p) \pmod{p}$. Then $s^4 \equiv 1 \pmod{p}$ and

$$s \equiv \pm i \pmod{p}. \tag{22}$$

Suppose $u(a, 1)$ is non-defective modulo p. Let $t = \alpha(a, 1, p)$. Since

$$u_0 \equiv u_t \equiv u_{2t} \equiv u_{3t} \equiv 0 \pmod{p},$$

we must have that

$$\alpha(a, 1, p) = (p + 1)/2 \text{ and } \mu(a, 1, p) = 4\alpha(a, 1, p) = 2(p + 1).$$

By (22) and the remark following Definition 2,

$$u_{p+1+k} \equiv (-1)u_k \pmod{p} \tag{23}$$

for $1 \leq k \leq p + 1$. By Lemma 6 (iii),

$$u_{p+1-k} \equiv (-1)^k u_k \pmod{p} \tag{24}$$

for $1 \leq k \leq p$. Let N_2 be the largest integer k such that there exist integers $n_1 n_2, \ldots, n_k$ for which $1 \leq n_i \leq (p - 1)/2$ and $u_{n_i} \not\equiv \pm u_{n_j}$ $\pmod{p}$ if $1 \leq i < j \leq (p - 1)/2$. Since $u_0 \equiv 0 \pmod{p}$, it follows from (23) and (24) that

$$N(a, 1, p) = 2N_2 + 1.$$

Thus, $u(a, 1)$ is non-defective modulo p if and only if $N_2 = (p - 1)/2$. This occurs only if

$$u_i \not\equiv \pm u_j \pmod{p}$$

for all integers i and j such that $1 \leq i, j \leq (p - 1)/2$ and $i \neq j$.

By Proposition 8, the derived sequence $\{u_{n(2r-1)}/u_{2r-1}\}$ is a LSFK

$u(v_{2r-1}, 1)$, where $2r - 1$ is a fixed odd integer, n varies over all non-negative integers, and $v(a, 1)$ is the LSSK associated with the LSFK $u(a, 1)$. We now consider the $(p - 1)/4$ LSFK's

$$\{u_{n(2r-1)}/u_{2r-1}\} = u(v_{2r-1}, 1) ,$$

where $1 \leq 2r - 1 \leq (p - 1)/2$. It is well-known that

$$v_{2r-1}^2 = Du_{2r-1}^2 - 4,$$

where D is the discriminant of $u(a, 1)$ (see, for example [7], page 317). Let D' be the discriminant of

$$\{u_{n(2r-1)}/u_{2r-1}\} = u(v_{2r-1}, 1).$$

Then

$$D' = v_{2r-1}^2 + 4 = (Du_{2r-1}^2 - 4) + 4 = Du_{2r-1}^2. \qquad (25$$

Since $(D/p) = -1$, $(D'/p) = -1$ also. By Lemma 7 (ii),

$$u_{2r_1-1} \not\equiv \pm u_{2r_2-1} \pmod{p} \qquad (26$$

for $1 \leq 2r_1 - 1 \leq 2r_2 - 1 \leq (p-1)/2$. Hence, by (25) and (26), the $(p-1)/4$ residues

$$v_{2r-1}^2 + 4 = Du_{2r-1}^2$$

are all distinct quadratic residues modulo p for $1 \leq 2r - 1 \leq (p - 1)/2$. However, by Proposition 9, there are exactly $(p - 1)/4$ quadratic residues c^2 such that $c^2 + 4$ is congruent to a quadratic non-residue modulo p.

We now note that 1 is a quadratic residue modulo p such that when it is added to 4, a quadratic non-residue, namely 5, is obtained. Thus, we must have that for some positive integer r such that $1 \leq 2r - 1 \leq (p - 1)/2$,

$$v_{2r-1}^2 \equiv 1 \ (\text{mod } p)$$

or

$$v_{2r-1} \equiv \pm 1 \ (\text{mod } p).$$

Thus, for this value of r, the LSFK

$$\{u_{n(2r-1)}/u_{2r-1}\} = u(\pm 1, 1).$$

We note that u(1, 1) is just the Fibonacci sequence. For the LSFK u($\pm$1, 1), the discriminant $D' = 5$ and $(5/p) = -1$. Hence, $\alpha(\pm 1, 1, p) \equiv 1 \ (\text{mod } 2)$, $\beta(\pm 1, 1, p) = 4$, and

$$u_{\alpha(\pm 1,1,p)-1} \equiv \pm i \ (\text{mod } p). \tag{27}$$

Let $k = \alpha(\pm 1, 1, p)$. Then $u_k(\pm 1, 1) \equiv 0 \ (\text{mod } p)$, and by the recursion relation defining u($\pm$1, 1) and (27),

$$u_{k-1}(\pm 1, 1) \equiv \pm i, \ u_{k-2}(\pm 1, 1) \equiv \pm i \ (\text{mod } p) \tag{28}$$

Also,

$$u_1(\pm 1, 1) \equiv 1, \ u_2(\pm 1, 1) \equiv \pm 1 \ (\text{mod } p). \tag{29}$$

We note that

$$u_{(2r-1)_i}(a, 1) \equiv u_{2r-1} \ u_i(\pm 1, 1) \not\equiv 0 \ (\text{mod } p) \tag{30}$$

for $1 \le i \le k - 1$. Thus,

$$(2r - 1)i \not\equiv (2r - 1)j \ (\text{mod } k) \tag{31}$$

for $1 \le i \le j \le k - 1$, since otherwise we would have

$$u_{(2r-1)(j-i)}(a, 1) \equiv 0 \ (\text{mod } p),$$

contrary to (30).

We note that

$$u_{2(2r-1)}(a, 1)/u_{2r-1}(a, 1) \equiv u_2(\pm 1, 1) \equiv \pm 1 \pmod p. \tag{32}$$

Since $s(a, 1, p) \equiv \pm i \pmod p$, it now follows that

$$u_{2(2r-1)}(a, 1) \equiv \pm u_{2r-1}(a, 1) \equiv i^{n_1} u_e(a, 1) \pmod p \tag{33}$$

where $0 \leq n_1 \leq 3$ and $1 \leq e \leq (p - 1)/2$. Similarly,

$$u_{(k-2)(2r-1)}(a, 1)/u_{2r-1}(a, 1) \equiv u_{k-2}(\pm 1, 1) \equiv \pm i \pmod p \tag{34}$$

and hence,

$$u_{(k-2)(2r-1)}(a, 1) \equiv \pm i u_{2r-1}(a, 1) \equiv i^{n_2} u_f(a, 1) \pmod p, \tag{35}$$

where $0 \leq n_2 \leq 3$ and $1 \leq f \leq (p - 1)/2$. Moreover,

$$u_{(k-1)(2r-1)}(a, 1)/u_{2r-1}(a, 1) \equiv u_{k-1}(\pm 1, 1) \equiv \pm i \pmod p \tag{36}$$

and thus,

$$u_{(k-1)(2r-1)}(a, 1) \equiv \pm i u_{2r-1}(a, 1) \equiv i^{n_3} u_g(a, 1) \pmod p, \tag{37}$$

where $0 \leq n_3 \leq 3$ and $1 \leq g \leq (p - 1)/2$. It follows from (31) that the integers $2r - 1$, e, f, and g are all distinct. Since $i^n = 1$, i, -1, or $-i$, it follows from congruences (33), (35), (37), and the pigeonhole principle that there exists a pair of residues u_{m_1} and u_{m_2} among the residues u_{2r-1}, u_e, u_f, and u_g such that

$$u_{m_1} \equiv \pm u_{m_2} \pmod p,$$

where $1 \leq m_1 < m_2 \leq (p - 1)/2$. It now follows that

$$N_2 \leq ((p - 1)/2) - 1$$

and hence,

$$N(a, 1, p) = 2N_2 + 1 \leq 2\,(((p - 1)/2) - 1) + 1 = p - 2 < p.$$

This contradicts the assumption that u(a, 1) is non-defective modulo p and the proof of part (1) is complete.

(2) This follows by inspection.

(3) Suppose $D = a^2 + 4 \equiv 0 \pmod{p}$. Then $a^2 \equiv -4$ and $a \equiv \pm 2i \pmod{p}$.

The rest follows by induction. □

4. CONCLUDING REMARKS

If $b \neq \pm 1$ and -b is not equal to a positive square, then by the Artin conjecture, it is highly unlikely that one can find an analogue of Theorems 1 and 2. The Artin conjecture states that if $r \neq \pm 1$ and r is not a square, then there exists an infinite number of primes p for which the integer r is a primitive root modulo p. Theorem 3 below considers the rank of apparition of p in u(a, b) when -b is a primitive root modulo p.

Theorem 3: Let p be a prime. Suppose the fixed integer -b is a primitive root modulo p. Then there exists a LSFK u(a, b) such that u(a, b) is non-defective modulo p.

Remark: Suppose the integer $-b \neq -1$ and is not equal to a square. If the Artin conjecture is correct, it follows from Theorem 3 that for an infinite number of primes p, there exists a LSFK u(a, b) such that u(a, b) has a complete system of residues modulo p.

Before proving Theorem 3, we need the following lemmas.

Lemma 10: Let p be a prime. Suppose b is a fixed integer such that -b is a primitive root modulo p. Then there exists a LSFK u(a, b) such that $(D/p) = -1$ and $\alpha(a, b, p)$ has a maximal value of $p + 1$.

Proof: This is proved in [9], pages 125-128. □

Lemma 11: Let u(a, b) be a LSFK. Let p be a prime such that $p \not| b$. Let $k = \alpha(a, b$

p). Let $s \equiv u_{k+1}$ (mod p) be the multiplier of u(a, b) modulo p. Then

$$s^2 \equiv (-b)^k \text{ (mod p)}.$$

Proof: This is proved in [9], pages 48-49. □

   We are now ready to prove Theorem 3.

Proof of Theorem 3: By Lemma 10, there exists a LSFK u(a, b) such that $\alpha(a, b, p)$
$= p + 1$. Let s be the multiplier of u(a, b) modulo p. Then by Lemma 11,

$$s^2 \equiv u_{p+2}^2 \equiv (-b)^{p+1} \equiv (-b)^{p-1} (-b)^2 \equiv (-b)^2 \text{ (mod p)}.$$

Thus, $s \equiv \pm b$ (mod p). If $s \equiv b$ (mod p), then by Proposition 2, the LSFK
u(-a, b) has $\alpha(-a, b, p) = p + 1$ and multiplier $s(-a, b, p) \equiv u_{p+2} \equiv -b$ (mod
p). Thus, there exists a LSFK u(a', b) with multiplier s(a', b, p) congruent
to a primitive root modulo p. Let $s_1 \equiv s(a', b, p)$ (mod p). By the remark
following Definition 2, the p - 1 terms

$$u_{n\alpha(a',b,p)+1} \equiv s_1^n u_1 \equiv s_1^n \text{ (mod p)},$$

are distinct modulo p for $0 \leq n \leq p - 2$. Adding the term $u_0 \equiv 0$ (mod p),
we see that u(a',b) has a complete system of residues modulo p. □

   For completeness, we add Theorem 4 which considers the case in which
b = 0.

Theorem 4: Let u(a, 0) be a LSFK. Let p be a prime. Then there exists a LSFK
   u(a, b) which is non-defective modulo p. In particular, the LSFK u(a, 0)
   is non-defective modulo p if and only if a is a primitive root modulo p

Proof: This follows immediately since $u_0 = 0$ and

$$u_n \equiv a^{n-1} \text{ (mod p)} \qquad \text{if } n \geq 1. \text{ □}$$

REFERENCES

[1] Bruckner, G. "Fibonacci Sequence Modulo a Prime p = 3 (mod 4)." *The Fibonacci Quarterly 8, No. 2* (1970): pp 217-220.

[2] Burr, S. A. "On Moduli for Which the Fibonacci Sequence Contains a Complete System of Residues." *The Fibonacci Quarterly 9, No. 4* (1971): pp 497-504.

[3] Lehmer, D. H. "An Extended Theory of Lucas' Functions." *Ann. Math. Second Series 31* (1930): pp 419-448.

[4] Shah, A. P. "Fibonacci Sequence Modulo m." *The Fibonacci Quarterly 6, No. 1* (1968): pp 139-141.

[5] Somer, L. "The Fibonacci Ratios F_{x+1}/F_x Modulo p." *The Fibonacci Quarterly 13, No. 4* (1975): pp 322-324.

[6] Somer, L. "Fibonacci-Like Groups and Periods of Fibonacci-Like Sequences." *The Fibonacci Quarterly 15, No. 1* (1977): pp 35-41.

[7] Somer, L. "The Divisibility Properties of Primary Lucas Recurrences with Respect to Primes." *The Fibonacci Quarterly 18, No. 4* (1980): pp 316-334.

[8] Somer, L. "Possible Periods of Primary Fibonacci-Like Sequences with Respect to a Fixed Odd Prime." *The Fibonacci Quarterly 20, No. 4* (1982): pp 311-333.

[9] Somer, L. "The Divisibility and Modular Properties of kth-order Linear Recurrences Over the Ring of Integers of an Algebraic Number Field with Respect to Prime Ideals". *Ph.D. Thesis. The University of Illinois at Urbana-Champaign*, 1985.

[10] Wyler, O. "On Second-Order Recurrences." *Amer. Math. Monthly 72* (1965): pp 500-506.

John R. Burke and Gerald E. Bergum

COVERING THE INTEGERS WITH LINEAR RECURRENCES

The idea of representing a given set as the union of subsets is quite common in mathematics. Sometimes the problem is very geometric in nature such as tiling the plane with a given geometric shape. Another common application of the idea of covering occurs in additive number theory. To illustrate this, let

$$Z_0 = \{0, 1, 2, \ldots \}.$$

Define a set $A \subset Z_0$ as a basis of order z if every $n \in Z_0$ can be represented as the sum of two not necessarily distinct elements of A. That is, $A \subset Z_0$ is a basis of order 2 if

$$\bigcup_{a_i \in A} (a_i + A) = Z_0.$$

This can be viewed as a covering problem.

As an example, let $A = \{\sum' 2^{2v}\} \cup \{\sum' 2^{2v+1}\} \cup \{0\}$ where $\sum'$ denotes a finite sum. Then A is a basis of order 2. It may also be observed that A is minimal in the sense that if any element of A is deleted, it is no longer a basis. Thus the family of sets $\{a_i + A\}_{a_i \in A}$ forms a minimal covering of Z_0 in a natural way.

In 1985, Simpson [1] studied coverings of the integers with arithmetic progressions. He defined regular coverings and disjoint coverings for arithmetic progressions. In the following we will consider these ideas as applied to linear recurrences.

Definition 1: A family of nth order linear recurring sequences is a regular covering of Z_0 if each $m \in Z_0$ occurs in at least one of the linear recurring sequences, and no proper subcollection has this property.

143

A. N. Philippou et al. (eds.), Applications of Fibonacci Numbers, 143–147.
© 1988 by Kluwer Academic Publishers.

Definition 2: A regular covering is a disjoint covering if no positive integer is contained in two or more of the linear recurring sequences.

The first question is of course, do such coverings exist? It should be noted that arithmetic progressions, at least the nonnegative terms, are the terms of a first order linear recurrence. Thus the question has been answered in the affirmative by Simpsons work.

Example 1: Consider the recurrence $U_{n+1} = U_{n+1}$ with $U_0 = 0$. Then $U_n = n$ and we have a "family" of one recurrence which covers Z_0.

For any $k>1$, Example 1 can be generalized to construct a family of k first order recurrences which is a disjoint covering. To see this, let k be any positive integer greater than one. For each $1 \leq i \leq k$, define $U_{(n+1)_i} = U_{n_i} + k$ as the linear recurrence with the initial condition $U_{0_i} = i$. These k first order linear recurrences are a disjoint covering of Z_0. When $k = 2$, $\{u_{m_1}\}$ is the set of positive odd integers while $\{u_{m_2}\}$ is the set of positive even integers.

Thus finding finite families that form disjoint covers is a rather simple matter. The next obvious question is whether or not we can find an infinite family of linear recurring sequences that form a disjoint covering of Z_0. An example will show that the answer is affirmative.

Example 2: Consider the recurrences $U_{(n+1)_t} = 2U_{n_t}$ with $U_{0_t} = t$ where $(2, t) = 1$ and $U_{0_0} = 0$. Since every $n \in Z_0$, $n \neq 0$, can be uniquely expressed as $n = 2^\delta t$ with $(2, t) = 1$, we have a disjoint cover which is clearly infinite.

The ideas involved in the two previous examples are exemplary of the general situation.

Theorem 1: Let k and n be arbitrary but fixed positive integers. There exists an nth order linear recurrence that generates a disjoint cover of Z_0 with k sequences.

<u>Proof</u>: Example 1 and its generalization covers the case $n = 1$, $k \geq 1$. We may therefore assume that $n \geq 2$. Consider any linear recurrence whose characteristic polynomial is of the form $(x-1)^2 p(x)$ where deg $(p(x)) = \delta \geq 0$ and $\delta+2 = n$. If the roots of $p(x)$ are $\alpha_1, \ldots, \alpha_\delta$ then the mth term of the recurrence can be expressed by

$$U_m = a_1 + na_2 + a_3\alpha_1^m + \ldots + a_{\delta+2}\alpha_\delta^m$$

Let $a_j = 0$ for $3 \leq j \leq \delta+2 = n$. By choosing $a_1 = j$, $j = 0, 1, \ldots, k-1$ with $a_2 = 1$ we generate k recurrence sequences that form a disjoint cover of Z_0.

<u>Corollary</u>: Let $n \geq 3$ and $k \geq 1$ be arbitrary but fixed integers. There exists infinitely many nth order recurrences, each of which will give rise to a disjoint cover of k sequences.

<u>Proof</u>: If $n \geq 3$ then there are infinitely many choices for the polynomial $p(x)$ in the proof of Theorem 1.

   In relation to the corollary of Theorem 1, there is still the question of how many recurrences generate a family of k-recurrence sequences forming a disjoint cover. For a first order recurrence, the recurrence of Example 1 (and the discussion that follows it) is the only one that works.

   For second order recurrences, there is only one homogeneous recurrence relation that will work. It is $U_{n+1} = 2U_n - U_{n-1}$. It should be noted that in this case, $U_n = a + bn$. To form k recurrence sequences forming a disjoint cover, choose $a_i = i$, $i = 0, 1, 2, \ldots, k-1$ and $b_i = k + i$.

   We now examine the case of an infinite family of recurrence sequences that forms a disjoint cover of Z_0. The model in this case is essentially that of Example 2.

<u>Theorem 2</u>: Any linear recurrence with a prime characteristic root can generate an infinite disjoint cover of Z_0.

<u>Proof</u>: Let p(x) be the characteristic polynomial of a recurrence with a root q, q
prime. Let α_i, i = 2, 3, . . . , m = deg(p(x)) be the other roots. Then using
the Binet formula,

$$U_n = aq^n + b_2(n)\alpha_2^n + b_3(n)\alpha_3^n + \ldots b_m(n)\alpha_m^n$$

where the functions $b_i(n)$ allow for the possibility of repeated roots.

By letting a run thru all values (a, q) = 1 together with a = 0 and
always letting $b_i(n)$ = 0 one obtains an infinite family of recurrence
sequences that form a disjoint cover for Z_0.

If there are no prime roots then the situation appears to be much more
complicated. It is still, however, an easy task to form regular coverings as
the following illustrates.

Consider the recurrence $U_{n+1} = U_n + U_{n-1}$. Start with initial conditions
$U_{0_0} = 1$, $U_{1_0} = 1$. Now choose U_{0_i} and U_{1_i} to be the least integers not yet
covered. By the choice of the initial conditions we have a regular cover.
Since the terms of each sequence grow geometrically, a simple density
argument shows that an infinite family of sequences is required.

<u>Example</u> 3: Consider the Fibonacci sequence. Then we get the sequences

1	1	2	3	5	8	13	21
4	6	10	16	26	42	68	110
7	9	16	25	41	66	107	173
11	12	23	35	58	93	151	244
14	15	29	44	73	117	190	307

. . .

. . .

Note that the covering is not disjoint and we hve been unable to find
such a disjoint covering or to prove that none exists.

The present problem is to determine when two recurrence sequences will overlap given only the recurrence relation and intitial conditions. This, in turn, leads to some exponential Diophantine equations. To solve these in general seems to be a hopeless task.

An alternative is to consider sequences generated by several different recurrences and when one can expect to create regular or disjoint covers. This also seems to lead to the same overlap problem.

REFERENCE

[1] Simpson, R.J., "Regular Coverings of The Integers" by Arithmetic Progressions
     Acta Arith. (1985): pp 145-152.

Andreas N. Philippou

RECURSIVE THEOREMS FOR SUCCESS RUNS AND RELIABILITY OF CONSECUTIVE-K-OUT-OF-N: F SYSTEMS

1. INTRODUCTION AND SUMMARY

Unless otherwise stated, in this paper k is a fixed positive integer, n_i $(1 \leq i \leq k)$ and n are non-negative integers as specified, p and x are real numbers in the intervals $(0, 1)$ and $(0, \infty)$, respectively, and $q = 1-p$. In a recent paper, Philippou and Makri [11] studied the length L_n $(n \geq 1)$ of the longest success run in n Bernoulli trials, deriving the probability function of L_n, its distribution function, and its factorial moments. In particular, they found that

$$P(L_n \leq k) = \frac{p^{n+1}}{q} \sum_{n_1, \ldots, n_{k+1}} \binom{n_1 + \cdots + n_{k+1}}{n_1, \ldots, n_{k+1}} (\frac{q}{p})^{n_1 + \cdots + n_{k+1}}, \quad 0 \leq k \leq n, \quad (1.1)$$

$$n_1 + 2n_2 + \cdots + (k+1)n_{k+1} = n+1$$

and

$$P(L_n \leq k) = \frac{p^{n+1}}{q} F_{n+2}^{(k+1)} (q/p), \quad 0 \leq k \leq n, \quad (1.2)$$

where $\{F_n^{(k)}(x)\}_{n=0}^{\infty}$ are the Fibnacci-type polynomials of order k [11, 15].

Although the calculation of $P(L_n \leq k)$, $0 \leq k \leq n$, by means of (1.1) is not difficult, it does involve finding first all the non-negative integer-solutions of $n_1 + 2n_2 + \cdots + (k+1)n_{k+1} = n+1$, $0 \leq k \leq n$, which may take a lot of computational effort, especially when k and n are large. On the other hand, if (1.2) is employed, one needs only to know the polynomials $F_{n+2}^{(k)}(x)$, $0 \leq k \leq n$, which are easily calculated from their definition. Even this simple task, however, may become quite cumbersome for large values of k and n.

Presently we derive simple formulas for $P(L_n < k)$, when $k \leq n \leq 4k+2$, and we give a recurrence on n for $P(L_n < k)$, if $n \geq 4k+3$ (see Theorem 2.1). We also consider the reliability $R(p; k, n)$ of a consecutive-k-out-of-n:F system [2-8, 10,

149

N. Philippou et al. (eds.), Applications of Fibonacci Numbers, 149–161.
1988 by Kluwer Academic Publishers.

17], and the number of Bernoulli trials N_k until the occurrence of a success run of length k [9, 11-16, 18, 19]. As a byproduct of Theorem 2.1, we obtain simple formulas for R(p; k, n) and $P(N_k=n)$, respectively, when $k \leq n \leq 4k+2$ and $k \leq n \leq 5k+3$ and recurrences on n for R(p; k, n) and $P(N_k=n)$, if $n \geq 4k+3$ and $n \geq 5k+4$ (see Corollary 2.1 and Theorem 3.1). The usefulness of Theorems 2.1 and 3.1 for computational purposes, especially when $k \leq n \leq 4k+2$ and $k \leq n \leq 5k+3$, respectively, is indicated by means of three comparative examples regarding Theorem 2.1 and its competing formulas. Finally, the applicability of Corollary 2.1 is illustrated by means of a fourth example.

2. A RECURSIVE THEOREM FOR LONGEST SUCCESS RUNS

In the present section we shall establish the following theorem, which simplifies considerably the calculation of $P(L_n < k)$, especially when $k \leq n \leq 4k+2$.

Theorem 2.1: Let L_n be a random variable denoting the length of the longest success run in n Bernoulli trials and set

$$Q_n = Q_n(k, p) = P(L_n < k), \quad n \geq 0.$$

Then

(a) $Q_n = 1-p^k-(n-k)qp^k, \quad k \leq n \leq 2k;$

(b) $Q_n = 1-p^k-(n-k)qp^k+(n-2k)qp^{2k}+\frac{1}{2}(n-2k)(n-2k-1)q^2p^{2k},$
       $2k+1 \leq n \leq 3k+1;$

(c) $Q_n = 1-p^k-(n-k)qp^k+(n-2k)qp^{2k}+\frac{1}{2}(n-2k)(n-2k-1)q^2p^{2k}$
       $-\frac{1}{2}(n-3k)(n-3k-1)q^2p^{3k}-\frac{1}{6}(n-3k)(n-3k-1)(n-3k-2)q^3p^{3k},$
       $3k+2 \leq n \leq 4k+2;$

(d) $Q_n = Q_{n-1}-qp^kQ_{n-1-k}, \quad n \geq 4k+3.$

We shall first establish the following Lemma.

Lemma 2.1: Let L_n and Q_n be as in Theorem 2.1. Then

$$Q_n=\begin{cases} 1, & 0 \leq n \leq k-1, \\ 1-p^k, & n = k, \\ Q_{n-1}-qp^kQ_{n-1-k}, & n \geq k+1. \end{cases}$$

Proof: Denote by $A_{n-i, k}$ the event that there are no k consecutive successes in the first n - i Bernoulli trials ($0 \leq i \leq k$), and let L_{n-i} denote the length of the longest success run in the same trials. Then

$$(L_{n-i} < k) = A_{n-i,\,k}, \quad 0 \le i \le k, \; n \ge k, \tag{2.1}$$

and

$$A_{n,\,k} = \bigcup_{i=1}^{k} (A_{n-i,\,k} \cap \underbrace{(FSS \ldots S)}_{i-1}), \quad n \ge k. \tag{2.2}$$

Therefore

$$(L_n < k) = \bigcup_{i=1}^{k} [(L_{n-i} < k) \cap \underbrace{(FSS \ldots S)}_{i-1}], \quad n \ge k, \tag{2.3}$$

which gives

$$Q_n = q \sum_{i=1}^{k} p^{i-1} Q_{n-i}, \quad n \ge k, \tag{2.4}$$

by means of the difinition of Q_n, the mutual exlusiveness of the events $[(L_{n-i} < k) \cap \underbrace{(FSS \ldots S)}_{i-1}]$, $1 \le i \le k$, and the independence of the trials with $P\{S\} = p$ and $P\{F\} = q$. Furthermore, the definition of Q_n trivially gives

$$Q_n = 1, \; 0 \le n \le k-1, \tag{2.5}$$

which implies

$$Q_k = q \sum_{i=1}^{k} p^{i-1} = 1-p^k, \text{ by means of (2.4).} \tag{2.6}$$

Finally, using (2.4) again, we get

$$Q_n - pQ_{n-1} = qQ_{n-1} - qp^k Q_{n-1-k}, \quad n \ge k+1,$$

so that

$$Q_n = Q_{n-1} - qp^k Q_{n-1-k}, \quad n \ge k+1. \tag{2.7}$$

Relations (2.5)-(2.7) establish the lemma.

Proof of Theorem 2.1: Part (a) is true for $n = k$, since $Q_k = 1-p^k$, by Lemma 2.1. We assume that it is true for $n = \delta$ ($k \le \delta \le 2k-1$). Then

$$Q_{\delta+1} = Q_\delta - qp^k Q_{\delta-k} = 1-p^k-(\delta-k)qp^k-qp^k = 1-p^k-(\delta+1-k)qp^k,$$

by Lemma 2.1 and the induction hypothesis, QED. Part (b) is true for $n = 2k + 1$, since

$$Q_{2k+1} = Q_{2k}-qp^kQ_k = 1-p^k-kqp^k-qp^k(1-p^k) = 1-p^k-(k+1)qp^k+ qp^{2k},$$

by Lemma 2.1 and Theorem 2.1(a). Assume that it is true for $n = \delta$ $(2k+1 \leq \delta \leq 3k)$. Then

$$\begin{aligned}
Q_{\delta+1} &= Q_\delta-qp^kQ_{\delta-k} \\
&= 1-p^k-(\delta-k)qp^k+(\delta-2k)qp^{2k}+\tfrac{1}{2}(\delta-2k)(\delta-2k-1)q^2p^{2k} \\
&\quad -qp^k[1-p^k-(\delta-2k)qp^k] \\
&=1-p^k-(\delta+1-k)qp^k+(\delta+1-2k)qp^{2k}+\tfrac{1}{2}(\delta+1-2k)(\delta-2k)q^2p^{2k},
\end{aligned}$$

by Lemma 2.1, the induction hypothesis and Theorem 2.1(a), QED.

Part (c) is true for $n = 3k + 2$, since

$$\begin{aligned}
Q_{3k+2} &= Q_{3k+1}-qp^kQ_{2k+1} \\
&= 1-p^k-(2k+1)qp^k+(k+1)qp^{2k}+\tfrac{1}{2}(k+1)kq^2p^{2k} \\
&\quad -qp^k[1-p^k-(k+1)qp^k+qp^{2k}] \\
&= 1-p^k-(2k+2)qp^k+(k+2)qp^{2k}+\tfrac{1}{2}(k+2)(k+1)q^2p^{2k}-q^2p^{3k},
\end{aligned}$$

by Lemma 2.1 and Theorem 2.1(b). Assume that it is true for $n = \delta$ $(3k+2 \leq \delta \leq 4k+1)$. Then

$$\begin{aligned}
Q_{\delta+1} &= Q_\delta-qp^kQ_{\delta-k} \\
&= 1-p^k-(\delta-k)qp^k+(\delta-2k)qp^{2k}+\tfrac{1}{2}(\delta-2k)(\delta-2k-1)q^2p^{2k} \\
&\quad -\tfrac{1}{2}(\delta-3k)(\delta-3k-1)q^2p^{3k}-\tfrac{1}{6}(\delta-3k)(\delta-3k-1)(\delta-3k-2)q^3p^{3k} \\
&\quad -qp^k[1-p^k-(\delta-2k)qp^k+(\delta-3k)qp^{2k}+\tfrac{1}{2}(\delta-3k)(\delta-3k-1)q^2p^{2k}] \\
&= 1-p^k-(\delta+1-k)qp^k+(\delta+1-2k)qp^{2k}+\tfrac{1}{2}(\delta+1-2k)(\delta-2k)q^2p^{2k} \\
&\quad -\tfrac{1}{2}(\delta+1-3k)(\delta-3k)q^2p^{3k}-\tfrac{1}{6}(\delta+1-3k)(\delta-3k)(\delta-3k-1)q^3p^{3k},
\end{aligned}$$

by Lemma 2.1, the induction hypothesis and Theorem 2.1(b), QED. We finally note that part (d) is trivially true, by Lemma 2.1, and this completes the proof of the theorem.

To Theorem 2.1 we have the following corollary which improves upon a result of Shanthikumar [17]. See, also, Philippou [10].

Corollary 2.1: Assume that the components of the consecutive k-out-of-n:F system are ordered linearly and function independently with probability p, and denote its reliability by $R(p; k, n)$. Then

(a) $R(p; k, n) = 1-q^k-(n-k)pq^k$, $k \leq n \leq 2k$;

(b) $R(p; k, n) = 1-q^k-(n-k)pq^k+(n-2k)pq^{2k}+\frac{1}{2}(n-2k)(n-2k-1)p^2q^{2k}$,
     $2k+1 \leq n \leq 3k+1$;

(c) $R(p; k, n) = 1-q^k-(n-k)pq^k+(n-2k)pq^{2k}+\frac{1}{2}(n-2k)(n-2k-1)p^2q^{2k}$
     $-\frac{1}{2}(n-3k)(n-3k-1)p^2q^{3k}-\frac{1}{6}(n-3k)(n-3k-1)(n-3k-2)p^3q^{3k}$,
     $3k+2 \leq n \leq 4k+2$;

(d) $R(p; k, n) = R(p; k, n-1)-R(p; k, n-1-k)pq^k$, $n \geq 4k+3$.

Proof: Denote by $\widetilde{L}_n$ the length of the longest failure run among the n components and let $Q_n(k, p)$ be as in Theorem 2.1. Then
$$R(p; k, n) = P(\widetilde{L}_n \leq k-1),$$
and
$$P(\widetilde{L}_n \leq k-1) = Q_n(k,q), \; n \geq 0.$$
Therefore
$$R(p; k, n) = Q_n(k, q), \; n \geq 0,$$
which establishes the corollary, by means of Theorem 2.1.

We proceed now to state and prove a recursive theorem for $P(N_k=n)$, $n \geq k$.

3. A RECURSIVE THEOREM FOR FIRST SUCCESS RUNS

Let N_k denote the number of Bernoulli trials until the first occurrence of a success run of length k. Shane [16], Turner [18], Philippou and Muwafi [12], and Philippou, Georghiou and Philippou [14, 15] derived alternative formulas for $P(N_k=n)$, $n \geq k$ $(k \geq 2)$. Implicit in [14] there was still another, which was explicitely stated and proved by Uppuluri and Patil [19]. We recall the following two [12, 15], which hold for $k=1$ too [11, 13].

$$P(N_k=n)=p^n \sum_{\substack{n_1, \ldots, n_k \geqslant}} \binom{n_1 + \ldots + n_k}{n_1, \ldots, n_k} (\tfrac{q}{p})^{n_1 + \ldots + n_k}, \; n \geq k; \qquad (3.1)$$

$$n_1+2n_2+ \ldots +kn_k=n-k$$

$$P(N_k=n) = p^n F_{n+1-k}^{(k)}(q/p), \; n \geq k. \qquad (3.2)$$

We also mention that Feller [9] derived two recurrences in common for the sequences of probabilities $P(N_k=n)$ and P(a success run occurs at the nth trial), and Philippou and Makri [11] obtained the following recurrence:

$$P(N_k=n) = \begin{cases} p^k, \; n = k, \\ qp^k, \; k+1 \leq n \leq 2k, \\ P(N_k=n-1)-qp^k P(N_k=n-1-k), \; n \geq 2k+1. \end{cases} \qquad (3.3)$$

For another version of (3.3), we refer to Aki, Kuboki and Hirano [1].

In this section we shall establish the following improved version of (3.3) as a corollary of Theorem 2.1.

Theorem 3.1: Let N_k be a random variable denoting the number of Bernoulli trials until the first occurrence of a success run of length k, and set

$$P_n = P_n(k, p) = P(N_k=n), \; n \geq k.$$

Then

(a) $P_n = p^k$, n=k;

(b) $P_n = qp^k$, $k+1 \leq n \leq 2k$;

(c) $P_n = qp^k[1-p^k-(n-2k-1)qp^k]$, $2k+1 \leq n \leq 3k+1$;

(d) $P_n = qp^k[1-p^k-(n-2k-1)qp^k+(n-3k-1)qp^{2k}+\tfrac{1}{2}(n-3k-1)$
$(n-3k-2)q^2p^{2k}]$, $3k+2 \leq n \leq 4k+2$;

(e) $P_n = qp^k[1-p^k-(n-2k-1)qp^k+(n-3k-1)qp^{2k}+\tfrac{1}{2}(n-3k-1)$
$(n-3k-2)q^2p^{2k}-\tfrac{1}{2}(n-4k-1)(n-4k-2)q^2p^{3k}-\tfrac{1}{6}(n-4k-1)$
$(n-4k-2)(n-4k-3)q^3p^{3k}]$, $4k+3 \leq n \leq 5k+3$;

(f) $P_n = P_{n-1}-qp^k P_{n-1-k}$, $n \geq 5k+4$.

Part (a) of the theorem is trivially true. Parts (b) and (c)-(f), respectively, are direct consequences of relation (2.5) and Theorem 2.1, by means of the following lemma, which provides a simple relationship for P_n and Q_n if $n \geq k+1$.

Lemma 3.1: Let Q_n and P_n be as in Theorems 2.1 and 2.2, respectively. Then

$$P_n = qp^k Q_{n-1-k}, \; n \geq k+1.$$

Proof: Since $Q_n = 1$ for $0 \leq n \leq k-1$, and $F_{n+2}^{(k)}(q/p) = \dfrac{q}{p^{n+1}}$ for $0 \leq n \leq k-1$, by the definition of $\{F_n^{(k)}(\cdot)\}_{n=0}^{\infty}$, we have

$$Q_n = \frac{p^{n+1}}{q} F_{n+2}^{(k)}(q/p), \; 0 \leq n \leq k-1.$$

This relation along with the definition of Q_n and (1.2) give

$$Q_n = \frac{p^{n+1}}{q} F_{n+2}^{(k)}(q/p), \; n \geq 0. \tag{3.4}$$

The lemma follows by comparison of (3.2) and (3.4).

4. COMPUTATIONAL EXAMPLES

In this section we illustrate the computational usefulness of Theorems 2.1 and 3.1 by means of three examples. Since both theorems, as well as their respective competing formulas, are of the same nature, it suffices to restrict our attention to one of them, say Theorem 2.1, in comparison to its competing formulas (1.1) and (1.2). We also give a fourth example which illustrates the applicability of Corollary 2.1.

Example 4.1: Assume that $k = 3$. Theorem 2.1 readily gives $Q_n(3, p)$ for $3 \leq n \leq 14$, and recursively for $n \geq 15$. In particular,

$$Q_8(3,p) = 1 - p^3 - 5qp^3 + 2qp^6 + q^2 p^6, \tag{4.1}$$

and

$$Q_{11}(3,p) = 1 - p^3 - 8qp^3 + 5qp^6 + 10q^2 p^6 - q^2 p^9. \tag{4.2}$$

Alternatively, if we use (1.1) we get

$$Q_8(3, p) = P(L_8 \leq 2) = \frac{p^9}{q} \sum_{\substack{n_1, n_2, n_3 \\ n_1+2n_2+3n_3=9}} \binom{n_1 + n_2 + n_3}{n_1, n_2, n_3} (\tfrac{q}{p})^{n_1+n_2+n_3} \tag{4.}$$

$$= \frac{p^9}{q} \left[(\tfrac{q}{p})^9 + 8(\tfrac{q}{p})^8 + (7+21)(\tfrac{q}{p})^7 + (30+20)(\tfrac{q}{p})^6 + (10+30+5)(\tfrac{q}{p})^5 \right.$$

$$\left. + (12+4)(\tfrac{q}{p})^4 + (\tfrac{q}{p})^3 \right]$$

$$= q^8 + 8pq^7 + 28p^2q^6 + 50p^3q^5 + 45p^4q^4 + 16p^5q^3 + p^6q^2,$$

and

$$Q_{11}(3,p) = P(L_{11} \leq 2) = \frac{p^{12}}{q} \sum_{\substack{n_1, n_2, n_3 \\ n_1+2n_2+3n_3=12}} \binom{n_1 + n_2 + n_3}{n_1, n_2, n_3} (\tfrac{q}{p})^{n_1+n_2+n_3} \tag{4.}$$

$$= \frac{p^{12}}{q} \left[(\tfrac{q}{p})^{12} + 11(\tfrac{q}{p})^{11} + (10+45)(\tfrac{q}{p})^{10} + (72+84)(\tfrac{q}{p})^9 \right.$$

$$+ (28+168+70)(\tfrac{q}{p})^8 + (105+140+21)(\tfrac{q}{p})^7$$

$$\left. + (20+90+30+1)(\tfrac{q}{p})^6 + (20+10)(\tfrac{q}{p})^5 + (\tfrac{q}{p})^4 \right]$$

$$= q^{11} + 11pq^{10} + 55p^2q^9 + 156p^3q^8 + 266p^4q^7 + 266p^5q^6$$

$$+ 141p^6q^5 + 30p^7q^4 + p^8q^3.$$

If we use (1.2), we get

$$Q_8(3, p) = P(L_8 \leq 2) = \frac{p^9}{q} F_{10}^{(3)}(q/p) \tag{4.}$$

$$= \frac{p^9}{q} \left[(\tfrac{q}{p})^9 + 8(\tfrac{q}{p})^8 + 28(\tfrac{q}{p})^7 + 50(\tfrac{q}{p})^6 + 45(\tfrac{q}{p})^5 + 16(\tfrac{q}{p})^4 + (\tfrac{q}{p})^3 \right]$$

$$= q^8 + 8pq^7 + 28p^2q^6 + 50p^3q^5 + 45p^4q^4 + 16p^5q^3 + p^6q^2,$$

and

$$Q_{11}(3, p) = P(L_{11} \leq 2) = \frac{p^{12}}{q} F_{13}^{(3)}(q/p) \tag{4.6}$$

$$= \frac{p^{12}}{q} \left[(\tfrac{q}{p})^{12} + 11(\tfrac{q}{p})^{11} + 55(\tfrac{q}{p})^{10} + 156(\tfrac{q}{p})^{9} + 266(\tfrac{q}{p})^{8} + 266(\tfrac{q}{p})^{7} \right.$$

$$\left. + 141(\tfrac{q}{p})^{6} + 30(\tfrac{q}{p})^{5} + (\tfrac{q}{p})^{4} \right]$$

$$= q^{11} + 11pq^{10} + 55p^{2}q^{9} + 156p^{3}q^{8} + 266p^{4}q^{7} + 266p^{5}q^{6}$$

$$+ 141p^{6}q^{5} + 30p^{7}q^{4} + p^{8}q^{3}.$$

Example 4.2: Assume that $k = 5$. Theorem 2.1 readily gives $Q_n(5, p)$ for $5 \leq n \leq 22$, and recursively for $n \geq 23$. In particular,

$$Q_{18}(5, p) = 1 - p^5 - 13qp^5 + 8qp^{10} + 28q^2p^{10} - 3q^2p^{15} - q^3p^{15}, \tag{4.7}$$

and

$$Q_{23}(5, p) = Q_{22}(5, p) - qp^5 Q_{17}(5, p) \tag{4.8}$$
$$= 1 - p^5 - 17qp^5 + 12qp^{10} + 66q^2p^{10} - 21q^2p^{15} - 35q^3p^{15}$$
$$\quad - qp^5[1 - p^5 - 12qp^5 + 7qp^{10} + 21q^2p^{10} - q^2p^{15}]$$
$$= 1 - p^5 - 18qp^5 + 13qp^{10} + 78q^2p^{10} - 28q^2p^{15} - 56q^3p^{15} + q^3p^{20}.$$

On the other hand, in order to calculate $Q_{18}(5, p)$ and $Q_{23}(5, p)$ by means of (1.1), we need to find, among other things, all non-negative integer-solutions of the equations

$$n_1 + 2n_2 + 3n_3 + 4n_4 + 5n_5 = 19 \quad \text{and} \quad n_1 + 2n_2 + 3n_3 + 4n_4 + 5n_5 = 24,$$

which require some computational effort. In order to calculate $Q_{18}(5, p)$ and $Q_{23}(5, p)$ by means of (1.2), we need $F_{20}^{(5)}(\cdot)$ and $F_{25}^{(5)}(\cdot)$. We do the first.

$$Q_{18}(5, p) = P(L_{18} \leq 4) = \frac{p^{19}}{q} F_{20}^{(5)}(q/p) \tag{4.9}$$

$$= \frac{p^{19}}{q} \left[(\tfrac{q}{p})^{19} + 18(\tfrac{q}{p})^{18} + 153(\tfrac{q}{p})^{17} + 816(\tfrac{q}{p})^{16} + 3060(\tfrac{q}{p})^{15} \right.$$

$$+8554\left(\tfrac{q}{p}\right)^{14}+18395\left(\tfrac{q}{p}\right)^{13}+30888\left(\tfrac{q}{p}\right)^{12}+40612\left(\tfrac{q}{p}\right)^{11}$$

$$+41470\left(\tfrac{q}{p}\right)^{10}+32211\left(\tfrac{q}{p}\right)^{9}+18320\left(\tfrac{q}{p}\right)^{8}+7140\left(\tfrac{q}{p}\right)^{7}$$

$$+1686\left(\tfrac{q}{p}\right)^{6}+185\left(\tfrac{q}{p}\right)^{5}+4\left(\tfrac{q}{p}\right)^{4}\Big]$$

$$= q^{18}+18pq^{17}+153p^2q^{16}+816p^3q^{15}+3060p^4q^{14}+8554p^5q^{13}$$

$$+18395p^6q^{12}+30888p^7q^{11}+40612p^8q^{10}+41470p^9q^9$$

$$+32211p^{10}q^8+18320p^{11}q^7+7140p^{12}q^6+1686p^{13}q^5$$

$$+185p^{14}q^4+4p^{15}q^3.$$

Example 4.3: Assume that $k = 20$. Theorem 2.1 readily gives $Q_n(20, p)$ for $20 \leq n \leq 82$, and recursively for $n \geq 83$. In particular,

$$Q_{60}(20, p) = 1-p^{20}-40qp^{20}+20qp^{40}+190q^2p^{40}, \qquad (4.10)$$

and

$$Q_{83}(20, p) = Q_{82}(20, p)-qp^kQ_{62}(20, p) \qquad (4.11)$$
$$= 1-p^{20}-62qp^{20}+42qp^{40}+861q^2p^{40}-231q^2p^{60}-1540q^3p^{60}$$
$$-qp^{20}[1-p^{20}-42qp^{20}+22qp^{40}+231q^2p^{40}-q^2p^{60}]$$
$$= 1-p^{20}-63qp^{20}+43qp^{40}+903q^2p^{40}-253q^3p^{60}$$
$$-1771q^3p^{60}+q^3p^{80}.$$

On the other hand, both formulas (1.1) and (1.2) do not appear to be applicable for calculating $Q_{60}(20, p)$ and $Q_{83}(20, p)$ without a lot of computational effort.

Example 4.4: (a) Assume that $k = 2$. Corollary 2.1 readily gives $R(p; 2, n)$ for $2 \leq n \leq 10$, and recursively for $n \geq 11$. In particular,

$$R(p; 2, 4) = 1-q^2-2pq^2, \qquad (4.12)$$

$$R(p; 2, 7) = 1-q^2-5pq^2+3pq^4+3p^2q^4, \tag{4.13}$$

and

$$R(p; 2, 9) = 1-q^2-7pq^2+5pq^4+10p^2q^4-3p^2q^6-p^3q^6, \tag{4.14}$$

which are consistent with computational examples of Bollinger and Salvia [5], Chiang and Niu [7], and Bollinger [4].

(b) Assume that $k = 3$. Corollary 2.1 readily gives $R(p; 3, n)$ for $3 \le n \le 14$, and recursively for $n \ge 15$. In particular,

$$R(p; 3, 10) = 1-q^3-7pq^3+4pq^6+6p^2q^6, \tag{4.15}$$

and

$$R(p; 3, 11) = 1-q^3-8pq^3+5pq^6+10p^2q^6-p^2q^9, \tag{4.16}$$

which are consistent with computational examples of Bollinger [3, 2].

(c) Assume that $k = 4$. Corollary 2.1 readily gives $R(p; 4, n)$ for $4 \le n \le 18$, and recursively for $n \ge 19$. In particular,

$$R(p; 4, 13) = 1-q^4-9pq^4+5pq^8+10p^2q^8, \tag{4.17}$$

$$R(p; 4, 17) = 1-q^4-13pq^4+9pq^8+36p^2q^8-10p^2q^{12}-10p^3q^{12}, \tag{4.18}$$

and

$$\begin{aligned}
R(p; 4, 20) &= R(p; 4, 18)-pq^4[R(p; 4, 14)+R(p; 4, 15)] \tag{4.19}\\
&= 1-q^4-14pq^4+10pq^8+45p^2q^8-15p^2q^{12}-20p^3q^{12}\\
&\quad -pq^4[1-q^4-10pq^4+6pq^8+15p^2q^8-p^2q^{12}+1-q^4-11pq^4\\
&\quad +7pq^8+21p^2q^8-3p^2q^{12}-p^3q^{12}]\\
&= 1-q^4-16pq^4+12pq^8+66p^2q^8-28p^2q^{12}-56q^3p^{12}\\
&\quad +4p^3q^{16}+p^4q^{16}.
\end{aligned}$$

e note in ending that (4.17) coincides with a computational example of Chen and wang [6], whereas (4.18) does not, since there is an error in the calculations of [6].

REFERENCES

[1]   Aki, S. "Kuboki, H., and Hirano, K. "On Discrete Distributions of Order k."
      Annals of the Institute of Statistical Mathemaics 36, No. 3 (1984): pp 431-440.

[2]   Bollinger, R. C. "Direct Computation for Consecutive-k-out-of-n:F Systems."
      IEEE Transactions on Reliability R-31, No. 5 (1982): pp. 444-446.

[3]   Bollinger, R. C. "Reliability and Runs of Ones." *Mathematics Magazine 57, No.
      1* (1984): pp. 34-37.

[4]   Bollinger, R. C. "Strict Consecutive-k-out-of-n:F Systems." *IEEE
      Transactions on Reliability R-34, No. 1* (1985): pp. 50-52.

[5]   Bollinger, R. C., and Salvia, A. A. "Consecutive-k-out-of-n:F Networks." *IEEE
      Transactions on Reliability R-31, No. 1* (1982): pp. 53-55.

[6]   Chen, R. W., and Hwang, F. K. "Failure Distributions of Consecutive-k-out-of-
      n:F Systems." *IEEE Transactions on Reliability R-34, No. 4* (1985): pp. 338-341.

[7]   Chiang, D. T., and Niu, S. C. "Reliability of Consecutive-k-out-of-n:F System.
      IEEE Transactions on Reliability R-30, No. 1 (1981): pp 87-89.

[8]   Derman C., Lieberman, G. J., and Ross, S. M. "On the Consecutive-k-out-of-n:F
      System." *IEEE Transactions on Reliability R-31, No. 1* (1982): pp 57-63.

[9]   Feller, W. *An Introduction to Probability Theory and Its Applications, Vol. I,
      3rd ed.* New York: Wiley, 1968.

[10] Phillippou, A. N. "Distributions and Fibonacci Polynomials of Order k, Longest
      Runs, and Reliability of Consecutive-k-out-of-n:F Systems." In *Fibonacci
      Numbers and Their Applications: Proceedings of the First International
      Conference on Fibonacci Numbers and Their Applications (Patras 1984),*
      pp. 203-227, edited by A.N. Philippou, G.E. Bergum, and A.F. Horadam.
      Dordrecht: D. Reidel Publishing Company, 1986.

[11] Philippou, A. N., and Makri, F. S. "Longest Success Runs and Fibonacci-Type
      Polynomials." *The Fibonacci Quarterly 23, No. 4* (1985): pp 338-346.

[12] Philippou, A. N., and Muwafi, A. A. "Waiting for the kth Consecutive Success
      and the Fibonacci Sequence of Order k." *The Fibonacci Quarterly 20, No. 1*
      (1982): pp 28-32.

[13] Philippou, A. N., Georghiou, C., and Philippou, G. N. "A Generalized Geometric
      Distribution and Some of Its Properties." *Statistics and Probability Letters 1,
      No. 4* (1983): pp 171-175.

4] Philippou, A. N., Georghiou, C., and Philippou, G. N.   "Fibonacci Polynomials of
    Order k, Multinomial Expansions and Probability." *International Journal of*
    Mathematics and Mathematical Sciences 6, No. 3 (1983): pp 545-550.

5] Philippou, A. N., Georghiou, C., and Philippou, G. N.   "Fibonacci-type
    Polynomials of Order k with Probability Applications." *The Fibonacci*
    Quarterly 23, No. 2 (1985): pp 100-105.

6] Shane, H. D. "A Fibonacci Probability Function." *The Fibonacci Quarterly 11,*
    No. 6 (1973): pp 517-522.

7] Shanthikumar, J. G. "Recursive Algorithm to Evaluate the Reliability of a
    Consecutive-k-out-of-n:F System." *IEEE Transactions on Reliability R-31, No.*
    5 (1982): pp 442-443.

8] Turner, S. J. "Probability via the nth Order Fibonacci-T Sequence." *The*
    Fibonacci Quarterly 17, No. 1 (1979): pp 23-28.

9] Uppuluri, V. R. R., and Patil, S. A. "Waiting Times and Generalized Fibonacci
    Sequences." *The Fibonacci Quarterly 21, No. 4* (1983): pp 342-349.

A. F. Horadam and A. G. Shannon

ASVELD'S POLYNOMIALS $p_j(n)$

1. INTRODUCTION

In this paper we generalise the results of Asveld [1] for G_n. His method and notation will be followed where practicable. Our main objective is to extend his polynomials $p_j(n)$ for generalised Fibonacci numbers.

For this purpose, we need to utilise the sequence $\{U_n\}$ defined recursively by

$$U_n = p\, U_{n-1} - q\, U_{n-2} \tag{1.1}$$

with initial conditions

$$U_0 = 1,\ U_1 = p \quad (U_{-1} = 0) \tag{1.1}$$

This choice of the initial conditions allows us to follow more closely Asveld's usage of F_n for Fibonacci numbers, by which his subscript is 1 less than the usual subscript for Fibonacci numbers. In this spirit we shall say that

$$U_n = F_n \text{ when } p = 1,\ q = -1 \tag{1.3}$$

The Binet form for U_n is

$$U_n = (\alpha^{n+1} - \beta^{n+1})/\Delta \tag{1.4}$$

where

$$\alpha = (p + \sqrt{p^2 - 4q})/2,\ \beta = (p - \sqrt{p^2 - 4q})/2 \tag{1.5}$$

are the roots of the characteristic equation of (1.1), namely,

$$\lambda^2 - p\lambda + q = 0 \tag{1.6}$$

A. N. Philippou et al. (eds.), Applications of Fibonacci Numbers, 163–176.
© 1988 by Kluwer Academic Publishers.

so that

$$\alpha + \beta = p, \quad \alpha - \beta = \sqrt{p^2 - 4q} = \Delta \tag{1.7}$$

When $p = 1$, $q = -1$, as in (1.3), the value of α in (1.5) becomes

$$\alpha' = \frac{1 + \sqrt{5}}{2} \quad \text{for Fibonacci numbers.} \tag{1.8}$$

2. THE GENERALISATION

Consider the difference equation

$$H_n = p \, H_{n-1} - q \, H_{n-2} + (p - q - 1) \sum_{j=0}^{k} \alpha_j \, n^j \tag{2.1}$$

in which $\alpha_j (j = 0, 1, \ldots, k)$ are rational or real, and for which

$$H_0 = b, \; H_1 = pb - qa \tag{2.2}$$

Thus, for Asveld's G_n we have

$$H_n = G_n \text{ when } p = 1, \, q = -1, \, b = 1, \, a = 0 \tag{2.3}$$

Now, the solution $H_n^{(h)}$ of the *homogeneous* part of (2.1) is

$$H_n^{(h)} = A\alpha^n + B\beta^n \tag{2.4}$$

As a *particular* solution (symbolically $H_n^{(p)}$ where the bracketed symbol p has nothing whatsoever to do with the p in (1.1)), try

$$H_n^{(p)} = \sum_{i=0}^{k} A_i n^i \tag{2.5}$$

The total solution will then be

$$H_n = H_n^{(h)} + H_n^{(p)} \tag{2.6}$$

We propose to establish to following solution of (2.1):

$$H_n = (b + \Lambda_k)U_n + (\lambda_k - qa)U_{n-1} - \sum_{j=0}^{k} \alpha_j P_j(n) \tag{2.7}$$

in which

$$\Lambda_k = \sum_{j=0}^{k} a_{oj}\, \alpha_j \tag{2.8}$$

$$\lambda_k = \sum_{j=1}^{k} \left(\sum_{i=1}^{j} a_{ij} \right) \alpha_j - (p-1) \sum_{j=0}^{k} a_{oj}\alpha_j \tag{2.9}$$

$$P_j(n) = \sum_{i=0}^{j} a_{ij}\, n^i \tag{2.10}$$

where the a_{ij} are to be defined below.

Thus, Λ_k, λ_k are linear combinations of α_j, and $P_j(n)$ is a polynomial of degree j $(0 \leq j \leq k)$.

3. PROOF OF THE SOLUTION

Substitution of (2.5) in (2.1) yields

$$\sum_{i=0}^{k} A_i n^i - \sum_{i=0}^{k} \left\{ \sum_{\delta=0}^{i} A_i \binom{i}{\delta} (-1)^{i-\delta}(p-2^{i-\delta}q)n^\delta \right\} - (p-q-1) \sum_{i=0}^{k} \alpha_i\, n^i = 0 \tag{3.1}$$

after a little simplification.

Hence

$$A_i - \sum_{m=i}^{k} \beta_{im}A_m - (p-q-1)\,\alpha_i = 0 \quad (0 \leq i \leq k) \tag{3.2}$$

where

$$\beta_{im} = \binom{m}{i} (-1)^{m-i} (p-2^{m-i}q) \quad (i \leq m) \tag{3.3}$$

In particular, from (3.3),

$$\beta_{ii} = p - q \tag{3.4}$$

Next, set

$$A_i = - \sum_{j=i}^{k} a_{ij}\, \alpha_j \tag{3.5}$$

so that (3.2) becomes

$$- \sum_{j=i}^{k} a_{ij}\, \alpha_j + \sum_{m=i}^{k} \beta_{im} \left(\sum_{\delta=m}^{k} a_{m\delta}\alpha_\delta \right) - (p-q-1)\,\alpha_i = 0 \tag{3.6}$$

Using (3.4), we find

$$a_{tt} = 1 \tag{3.}$$

$$a_{ij} = -\sum_{m=i+1}^{j} \beta_{im} \, a_{mj} \quad (i < j) \tag{3.}$$

Substitution of (3.5) in (2.5) leads to

$$H_n^{(p)} = -\sum_{j=0}^{k} \alpha_j \left[\sum_{i=0}^{j} a_{ij} \, n^i \right] \tag{3.}$$

Use of the initial conditions (2.2) with (2.4) and (2.6) gives

$$H_0^{(h)} = A + B = b - H_0^{(p)} \tag{3.1}$$

and

$$H_1^{(h)} = A\alpha + B\beta = pb - qa - H_1^{(p)} \tag{3.1}$$

Solving (3.10) and (3.11) with the aid of (1.7), we obtain

$$A = \{(b - H_0^{(p)})\alpha + pH_0^{(p)} - H_1^{(p)} - qa\}/\triangle \tag{3.1}$$

and

$$B = \{-(b - H_0^{(p)})\beta - pH_0^{(p)} + H_1^{(p)} + qa\}/\triangle \tag{3.1}$$

Hence, by (1.4), (2.4), (2.6), (3.12) and (3.13)

$$H_n = (b - H_0^{(p)})U_n + (pH_0^{(p)} - H_1^{(p)} - qa)U_{n-1} + H_n^{(p)} \tag{3.1}$$

This is equivalent to (2.7) - (2.10), on referring to (3.9).

When $p = 1$, $q = -1$, $a = 0$, $b = 1$, Asveld's solution for G_n ensues (cf. (1.3) (2.3)).

4. THE POLYNOMIALS $P_j(N)$

The polynomials $P_j(n)$ for $j = 0,1,2, \ldots ,6$ are listed below in Table 1.

When $p = 1$, $q = -1$, $a = 0$, $b = 1$ we have our $P_j(n) =$ Asveld's $p_j(n)$.

Putting $p = 1$, $q = -1$ we derive the two infinite sequences (call them $\Lambda_{1,-1}$ and $\lambda_{1,-1}$) given by Asveld [1]:

$\Lambda_{1,-1}$:	1	3	13	81	673	6993	87193	1268361	21086113	. . . (4.1)
$\lambda_{1,-1}$:	0	1	7	49	415	4321	53887	783889	13031935	. . . (4.2)
	$p_0(n)$	$p_1(n)$	$p_2(n)$	$p_3(n)$	$p_4(n)$	$p_5(n)$	$p_6(n)$	$p_7(n)$	$p_8(n)$ . . .	

The numbers in the sequence (4.1) are seen to be the constants in $p_0(n)$, $p_1(n)$, $p_2(n)$, . . . , $p_8(n)$, . . .

The numbers in the sequence (4.2) are the sums of the coefficients of the powers of n in $p_1(n)$, $p_2(n)$, . . . , $p_8(n)$, . . .

As Asveld remarks, sequences (4.1) and (4.2) do not occur in [2].

Coming to the special case of (1.1) when $p = 2$, $q = -1$, so that we have the *Pell numbers* instead of the Fibonacci numbers, and writing $\Lambda_{2,-1}$ and $\lambda_{2,-1}$ for the two infinite sequences, we find

$\Lambda_{2,-1}$:	1	4	26	250	3206	51394	988646	. . .	(4.3)
$\lambda_{2,-1}$:	0	1	9	91	1173	18811	361869	. . .	(4.4)
	$p_0'(n)$	$p_1'(n)$	$p_2'(n)$	$p_3'(n)$	$p_4'(n)$	$p_5'(n)$	$p_6'(n)$	. . .	

in which the $p_j'(n)$ are the $P_j(n)$ for Pell numbers.

Again, these sequences (4.3) and (4.4) are not listed in [2].

Proceeding as for (4.1)-(4.4) we can establish pairs of infinite sequences $\Lambda_{p,-1}$ and $\lambda_{p,-1}$ which we conjecture might be unlisted in [2].

Some interesting limiting ratios may be spotted between pairs of elements in the sequences.

Let us denote by $\Lambda_\delta^{(p)}$ and $\lambda_\delta^{(p)}$ the elements in the δth positions in $\Lambda_{p,-1}$ and $\lambda_{p,-1}$ respectively.

Consider

$$\lim_{\delta \to \infty} \left(\frac{\Lambda_\delta^{(p)}}{\lambda_\delta^{(p)}} \right) \tag{4.5}$$

where

$$\Lambda_\delta^{(p)} = P_\delta(0) \tag{4.6}$$

and

$$\lambda_\delta^{(p)} = P_\delta(1) - P_\delta(0) \tag{4.7}$$

on referring to (4.1) - (4.4) and (2.10).

Data obtained from machine computation suggest that for $p = 1$, 2 and 4, the values of the limiting ratios (4.5) are $\frac{1+\sqrt{5}}{2}$ ($=\alpha'$ in (1.8)), $1+\sqrt{3}$ and $2+2\sqrt{2}$ respectively.   When p = 3, the limiting value is about 3.7913 for which an expression involving a square root, if it exists, is difficult to ascertain.

However, more theory at our disposal in the next section will enable us to discover, in Theorem 5, a general form of (4.5).   This will illumine the hidden pattern uniting the particular values of (4.5) given in the preceding paragraph. See (5.12).

S5. THE NUMBERS a_{nm}

Next, we investigate relationships among the numbers a_{nm} occurring in the expression (2.10) for $P_j(n)$. Tables 2 and 3 display some values of the a_{nm} for the Fibonacci and Pell numbers, respectively.

Firstly, we obtain

<u>Lemma 1</u>: $\dfrac{a_{n,n+1}}{a_{n-1,n}} = \dfrac{\beta_{n,n+1}}{\beta_{n-1,n}} = \dfrac{n+1}{n}$

<u>Proof</u>:   From (3.7) and (3.8),

$$a_{n,n+1} = -\beta_{n,n+1}$$
$$= (n+1)\,(p-2q) \qquad \text{by (3.3)}$$

The Lemma follows immediately.

<u>Lemma 2</u>: $n\,a_{nm} = m\,a_{n-1,m-1}$

<u>Proof</u>: Now

$$-a_{nm} = \sum_{t=n+1}^{m} \beta_{nt}\,a_{tm} \qquad \text{... (i)}$$

so

$$-a_{n-1,m-1} = \sum_{t=n}^{m-1} \beta_{n-1,t}\,a_{t,m-1} \qquad \text{... (ii)}$$

Compare the m-n ratios

$$\frac{\beta_{nt}\,a_{tm}}{\beta_{n-1,t}\,a_{t,m-1}} \qquad \begin{array}{l} (t = n+1, \ldots, m) \\ (t = n, \ldots, m-1) \end{array}$$

By Lemma 1, their values are successively

$$\frac{n+1}{n} \cdot \frac{m}{n+1} , \frac{n+2}{n} \cdot \frac{m}{n+2} , \frac{n+3}{n} \cdot \frac{m}{n+3} , \ldots , \frac{m-1}{n} \cdot \frac{m}{m-1} , \frac{m}{n} \cdot \frac{1}{1}$$

$$= \frac{m}{n} , \quad \frac{m}{n} , \quad \frac{m}{n} , \ldots , \quad \frac{m}{n} , \quad \frac{m}{n}$$

Hence, the Lemma is proved.

Successive applications of Lemma 2 lead to

<u>Theorem 1</u>: $a_{nm} = \binom{m}{n} a_{0,m-n}$

<u>Proof</u>: $a_{nm} = \frac{m}{n} a_{n-1,m-1}$ by Lemma 2

$\qquad = \frac{m}{n} \cdot \frac{m-1}{n-1} a_{n-2,m-2}$ by Lemma 2 again

$\qquad = \ldots \ldots \ldots \ldots \ldots$

$\qquad = \binom{m}{n} a_{0,m-n}$ by Lemma 2 repeatedly.

For example,

$$a_{26} = 10095 = 15 \times 673 = \binom{6}{2} a_{04} \qquad \text{for Fibonacci numbers}$$
$$= 48090 = 15 \times 3206 = \binom{6}{2} a_{04} \qquad \text{for Pell numbers}$$

An instance of Theorem 1 which we will need to use is the case n=1, namely

$$a_{1m} = m \, a_{0,m-1} \tag{5.1}$$

Other properties of the a_{nm} may now be established. Let us write

$$\lim_{m \to \infty} \left[\frac{a_{0m}}{a_{1m}} \right] = k$$

where k is a constant to be determined.

Firstly, we prove

<u>Theorem 2</u>: $\lim\limits_{m \to \infty} \left[\dfrac{a_{0m}}{a_{0,m-1}} - mk \right] = 0.$

<u>Proof</u>: $\lim\limits_{m \to \infty} \left[\dfrac{a_{0m}}{a_{0,m-1}} - mk \right] = \lim\limits_{m \to \infty} \left[\dfrac{a_{0m}}{a_{1m}} \cdot \dfrac{a_{1m}}{a_{0,m-1}} - mk \right]$

$$= \lim_{m \to \infty} mk - \lim_{m \to \infty} mk \qquad \text{by (5.1) and (5.2)}$$

$$= 0$$

Next, we derive

Theorem 3: $\qquad \lim\limits_{m \to \infty} \left(\dfrac{a_{0m}}{a_{nm}}\right) = n! \, k^n$

Proof: Induction on n is used.

When n = 0, the theorem is trivially true. When n = 1, the theorem holds, by (5.2). Suppose the theorem is valid for n = N, that is, assume

$$\lim\limits_{m \to \infty} \left(\dfrac{a_{0m}}{a_{Nm}}\right) = N! \, k^N \qquad \cdots\cdots\cdots\cdots (i)$$

Then, n = N+1 gives

$$\lim\limits_{m \to \infty} \left(\dfrac{a_{0m}}{a_{N+1,m}}\right) = \lim\limits_{m \to \infty} \left(\dfrac{a_{0m}}{a_{Nm}} \cdot \dfrac{a_{Nm}}{a_{N+1,m}}\right)$$

$$= N! \, k^N \lim\limits_{m \to \infty} \left(\dfrac{\binom{m}{N} \, a_{0,m-N}}{\binom{m}{N+1} \, a_{0,m-N-1}}\right) \text{ by (i) and Theorem 1}$$

$$= N! \, k^N \lim\limits_{m \to \infty} \left(\dfrac{N+1}{m-N} \cdot (m-N)k\right) \text{ by Theorem 2}$$

$$= (N+1)! \, k^{N+1}$$

The truth of the theorem is thus demonstrated.

It is now possible to determine a general numerical expression for k in (5.2) in terms of p and q.

From (3.8)

$$\frac{a_{0m}}{a_{1m}} = -\beta_{01}\frac{a_{1m}}{a_{1m}} - \beta_{02}\frac{a_{2m}}{a_{1m}} - \beta_{03}\frac{a_{3m}}{a_{1m}} - \cdots\cdots\cdots\cdots -\beta_{0m}\frac{a_{mm}}{a_{1m}}$$

$$= -\beta_{01}\frac{a_{1m}}{a_{0m}} \cdot \frac{a_{0m}}{a_{1m}} - \beta_{02}\frac{a_{2m}}{a_{0m}} \cdot \frac{a_{0m}}{a_{1m}} - \beta_{03}\frac{a_{3m}}{a_{0m}} \cdot \frac{a_{0m}}{a_{1m}} - \cdots$$

$$-\beta_{0m}\frac{a_{mm}}{a_{0m}} \cdot \frac{a_{0m}}{a_{1m}}$$

Divide throughout by $\dfrac{a_{0m}}{a_{1m}} \ (\neq 0)$. Let $m \to \infty$. Accordingly

$$1 = -\beta_{01} \lim\limits_{m \to \infty} \left(\frac{a_{1m}}{a_{0m}}\right) - \beta_{02} \lim\limits_{m \to \infty} \left(\frac{a_{2m}}{a_{0m}}\right) - \beta_{03} \lim\limits_{m \to \infty} \left(\frac{a_{3m}}{a_{0m}}\right) - \cdots$$

$$= (p-2q)\tfrac{1}{k} - (p-2^2q)\cdot\frac{1}{2!k^2} + (p-2^3q)\cdot\frac{1}{3!k^3} + \ldots \text{ by (3.3) and}$$
$$\text{Theorem 3}$$

$$= p\left\{\tfrac{1}{k} - \frac{1}{2!k^2} + \frac{1}{3!k^3} - \frac{1}{4!k^4} + \ldots\right\}$$

$$- q\left\{\tfrac{2}{k} - \tfrac{1}{2!}(\tfrac{2}{k})^2 + \tfrac{1}{3!}(\tfrac{2}{k})^3 - \tfrac{1}{4!}(\tfrac{2}{k})^4 + \ldots\right\}$$

$$= p(1-e^{-1/k}) - q(1-e^{-2/k}) \qquad \ldots\ldots\ldots\text{(ii)}$$

Write

$$t = e^{-1/k} \tag{5.3}$$

Then (ii) becomes

$$qt^2 - pt + p-1-q = 0 \tag{5.4}$$

Roots of (5.4) are

$$t = \left[p \pm \sqrt{p^2 - 4q(p-1-q)}\right]/2q \tag{5.5}$$

Since, by (5.3),

$$k = - \frac{1}{\log t} \tag{5.6}$$

only the positive root of (5.5) will concern us. Combine (5.2), (5.5) and (5.6).

Thus, we have shown that

Theorem 4: $$\lim_{m \to \infty}\left|\frac{a_{0m}}{a_{1m}}\right| = -1/\log[(p \pm \sqrt{p^2 - 4q(p-1-q)}/2q] \quad (>0)$$

(in which the negative root is ignored).

Putting $p = 1$, $q = -1$ in (5.5), (5.6) and Theorem 4, and using (1.8), we obtain

$$k = \frac{1}{\log \alpha'} = 2.078 \ldots \qquad \text{for Fibonacci numbers.} \tag{5.7}$$

Similarly, $p = 2$, $q = -1$ yield

$$k = 1/\log\left(\frac{1+\sqrt{3}}{2}\right) = 3.206 \ldots \quad \text{for Pell numbers.} \tag{5.8}$$

Computation discloses that the ratio $\frac{a_{0m}}{a_{1m}}$ approaches these limiting values

(5.7) and (5.8) rather rapidly for quite small values of m, e.g. $\frac{a_{08}}{a_{18}} =$ 2.0780867 . . . and $\frac{a_{08}}{a_{18}} = 3.2061007$ . . . in the Fibonacci and Pell cases, respectively.

We are now in a position to evaluate the limit given in (4.5). From (4.5), (4.6) and (4.7), we deduce that

$$\frac{\lambda_\delta^{(p)}}{\Lambda_\delta^{(p)}} = (a_{1\delta} + a_{1\delta} + a_{3\delta} + \ldots + a_{\delta\delta})/a_{0\delta} \tag{5.9}$$

Hence,

$$\lim_{\delta \to \infty} \left(\frac{\lambda_\delta^{(p)}}{\Lambda_\delta^{(p)}}\right) = \lim_{\delta \to \infty} \left(\frac{a_{1\delta}}{a_{0\delta}}\right) + \lim_{\delta \to \infty} \left(\frac{a_{2\delta}}{a_{0\delta}}\right) + \lim_{\delta \to \infty} \left(\frac{a_{3\delta}}{a_{0\delta}}\right) + \ldots$$

$$= \frac{1}{k} + \frac{1}{2!k^2} + \frac{1}{3!k^3} + \ldots \qquad \text{by Theorem 3}$$

$$= e^{1/k} - 1$$

That is,

<u>Theorem 5</u>: $$\lim_{\delta \to \infty} \left(\frac{\Lambda_\delta^{(p)}}{\lambda_\delta^{(p)}}\right) = (e^{1/k} - 1)^{-1}$$

If $p = 1$, $q = -1$ in (5.3) and (5.5), then $e^{1/k} - 1 = \frac{1+\sqrt{5}}{2} - 1 = \alpha' - 1$ by (1.8), giving

$$\lim_{\delta \to \infty} \left(\frac{\Lambda_\delta^{(1)}}{\lambda_\delta^{(1)}}\right) = \alpha' \qquad \text{for Fibonacci numbers.} \tag{5.10}$$

On the other hand, $p = 2$, $q = -1$ lead to

$$e^{1/k} - 1 = \frac{\sqrt{3}+1}{2} - 1 \text{ so that}$$

$$\lim_{\delta \to \infty} \left(\frac{\Lambda_\delta^{(2)}}{\lambda_\delta^{(2)}}\right) = 1 + \sqrt{3} \qquad \text{for Pell numbers.} \tag{5.11}$$

Results (5.10) and (5.11) confirm the computer-obtained information detailed in the paragraph following (4.7). Some other values of p and q give

p	q	t	Λ/λ
3	-1	$(\sqrt{21}-3)/2$	$(3+\sqrt{21})/2$
4	-1	$2\sqrt{2} - 2$	$2 + 2\sqrt{2}$
2	-2	$\dfrac{\sqrt{7} - 1}{2}$	$2 + \sqrt{7}$

$$(5.12)$$

One final remark on the a_{nm} may be offered. It relates to Theorem 2. Now

$$\frac{a_{0,m}}{a_{0,m-1}} = \frac{m}{m+1} \cdot \frac{a_{1,m+1}}{a_{1m}} \qquad \text{by Theorem 1} \qquad (5.13)$$

$$= \frac{m}{m+2} \cdot \frac{a_{2,m+2}}{a_{2,m+1}} \qquad \text{by Theorem 1 again}$$

$$= \cdots \cdots \cdots$$

$$= \frac{m}{m+r} \cdot \frac{a_{r,m+r}}{a_{r,m+r-1}} \qquad \text{by Theorem 1 used repeatedly.}$$

Proceeding to the limits as $m \to \infty$, we see that

$$\lim_{m \to \infty} \left[\frac{a_{0,m}}{a_{0,m-1}} \right] = \lim_{m \to \infty} \left[\frac{a_{1,m+1}}{a_{1m}} \right] = \lim_{m \to \infty} \left[\frac{a_{2,m+2}}{a_{2,m+1}} \right] = \cdots = \lim_{m \to \infty} \left[\frac{a_{r,m+r}}{a_{r,m+r-1}} \right] \qquad (5.14)$$

Starting out with the modest hope of generalising Asveld's work, we have been pleasantly surprised by the richness and variety of the consequences of our investigation. It has been a very satisfying experience.

TABLE 1. THE POLYNOMIALS $P_j(n)$ $(j=0, 1, 2, \ldots, 6)$

$P_0(n) = a_{00} = 1$

$P_1(n) = \displaystyle\sum_{i=0}^{1} a_{i1}n^i = a_{01} + a_{11}n = p-2q + n$

$P_2(n) = \displaystyle\sum_{i=0}^{2} a_{i2}n^i = a_{02}+a_{12}n+a_{22}n^2 = 2(p-2q)^2-(p-2^2q) + 2(p-2q)n + n^2$

$P_3(n) = \displaystyle\sum_{i=0}^{3} a_{i3}n^i = a_{03}+a_{13}n+a_{23}n^2+a_{33}n^3$
$$= 6(p-2q)^3-6(p-2q)(p-2^2q)+(p-2^3q)+[6(p-2q)^2-3(p-2^2q)]n+3(p-2q)n^2+n^3$$

$P_4(n) = \displaystyle\sum_{i=0}^{4} a_{i4}n^i = a_{04}+a_{14}n+a_{24}n^2+a_{34}n^3+a_{44}n^4$
$$= 24(p-2q)^4-36(p-2q)^2(p-2^2q)+6(p-2^2q)^2+8(p-2q)(p-2^3q)-(p-2^4q)$$
$$+ [24(p-2q)^3-24(p-2q)(p-2^2q)+4(p-2^3q]n+[12(p-2q)^2-6(p-2^2q)]n^2$$
$$+4(p-2q)n^3+n^4$$

$P_5(n) = \displaystyle\sum_{i=0}^{5} a_{i5}n^i = a_{05}+a_{15}n+a_{25}n^2+a_{35}n^3+a_{45}n^4+a_{55}n^5$
$$= [120(p-2q)^5-240(p-2q)^3(p-2^2q)+90(p-2q)(p-2^2q)^2+60(p-2q)^2(p-2^3q)$$
$$- 20(p-2^2q)(p-2^3q) - 10(p-2q)(p-2^4q)+(p-2^5q)]$$
$$+ [120(p-2q)^4-180(p-2q)^2(p-2^2q)+30(p-2^2q)^2+40(p-2q)(p-2^3q)-5(p-2^4q)]n$$
$$+ [60(p-2q)^3-60(p-2q)(p-2^2q)+10(p-2^3q)]n^2+[20(p-2q)^2-10(p-2^2q)]n^3$$
$$+ 5(p-2q)n^4+n^5$$

$P_6(n) = \displaystyle\sum_{i=0}^{6} a_{i6}n^i = a_{06}+a_{16}n+a_{26}n^2+a_{36}n^3+a_{46}n^4+a_{56}n^5+a_{66}n^6$
$$= [720(p-2q)^6-1800(p-2q)^4(p-2^2q)+1080(p-2q)^2(p-2^2q)^2$$
$$+ 480(p-2q)^3(p-2^3q)-360(p-2q)(p-2^2q)(p-2^3q)-90(p-2q)^2(p-2^4q)$$
$$+ 12(p-2q)(p-2^5q)-90(p-2^2q)^3+20(p-2^3q)^2-30(p-2^2q)(p-2^4q)-(p-2^6q)]$$
$$+ [720(p-2q)^5-1440(p-2q)^3(p-2^2q)+540(p-2q)(p-2^2q)^2$$
$$+360(p-2q)^2(p-2^3q)-120(p-2^2q)(p-2^3q)-60(p-2q)(p-2^4q)+6(p-2^5q)]n$$
$$+ [360(p-2q)^4-540(p-2q)^2(p-2^2q)+90(p-2^2q)^2+120(p-2q)(p-2^3q)$$
$$-15(p-2^4q)]n^2 + [120(p-2q)^3-120(p-2q)(p-2^2q)+20(p-2^3q)]n^3+[30(p-2q)^2$$
$$-15(p-2^2q)]n^4+6(p-2q)n^5+n^6$$

TABLE 2. a_{nm} FOR FIBONACCI NUMBERS

$n \backslash m$	0	1	2	3	4	5	6	7	. . . .
0	1	3	13	81	673	6993	87193	1268361	
1		1	6	39	324	3365	41958	610351	
2			1	9	78	810	10095	146853	
3				1	12	130	1620	23555	
4					1	15	195	2835	
5						1	18	273	
6							1	21	
7								1	
.									
.									
.									

TABLE 3. a_{nm} FOR PELL NUMBERS

$n \backslash m$	0	1	2	3	4	5	6	7	. . . .
0	1	4	26	250	3206	51394	988646	221887890	
1		1	8	78	1000	16030	308364	6920522	
2			1	12	156	2500	48090	1079274	
3				1	16	260	5000	112210	
4					1	20	390	8750	
5						1	24	546	
6							1	28	
7								1	
.									
.									
.									

REFERENCES

[1] Asveld, P. R. J. "A Family of Fibonacci-Like Sequences." *The Fibonacci Quarterly 25, No. 1* (1987): pp 81-83.

[2] Sloane, N. J. A. "A Handbook of Integer Sequences." *New York: Academic Press,* 1973.

Joseph Arkin and Gerald Bergum

MORE ON THE PROBLEM OF DIOPHANTUS

Recently, a number of articles have appeared in the literature which deal with finding a set of four numbers such that the product of any two different members when increased by n is a perfect square, [1] to [3] and [6] to [9].

The case n = 1 is credited to Fermat, [5], where he shows that {1, 3, 8, 120} have the designated property. In [3], Davenport and Baker show that a fifth integer x cannot be added to the set and still maintain the same property unless it is already one of the members of the set. In [6], Hoggatt and Bergum found an infinite family of four element sets where each set has the given property with n = 1. Three of the numbers in each set are in fact Fibonacci numbers.

In this paper, using only high school algebra, we determine a rule for finding five numbers, not all integers, such that the product of any two when increased by 1 is a square. We also find, by using new formulas consisting of the Fibonacci numbers, that we are able to solve problems similar to those considered by Fermat, Jones, Hoggatt and Bergum.

It is easy to verify that the four equations

$$A = x/2 - 1 \qquad (1)$$
$$B = x/2 + 1 \qquad (2)$$
$$C = 2x \qquad (3)$$
$$D = 2x^3 - 2x \qquad (4)$$

for any value of x give us a solution of the problem of Fermat, [5]. When x = 4, we get the solution given by Fermat. Furthermore, the polynomials (1) - (4) are not the polynomials given in [8] so the solutions are different and there are solutions given in [6] that are not given for any even value of x in (1) - (4) and conversely. Hence, our method generates new solutions to Fermat's Problem.

To add a fifth member to (1) - (4), we proceed as follows.

177

A. N. Philippou et al. (eds.), Applications of Fibonacci Numbers, 177–181.
© 1988 by Kluwer Academic Publishers.

J. ARKIN AND G. E. BERGUM

Let

$$E = (4DW + D^3)/4W^2 \tag{5}$$

then

$$ED+1 = ((D^2+2W)/2W)^2. \tag{6}$$

Now, we wish to find a value for W which will make AE+1, BE+1 and CE+1 squares. To do this we multiply (3) by (5) and substitute (4) for D to obtain

$$CE+1 = \frac{(16x(x^3-x)W+16(x^3-x)^3x+4W^2)}{4W^2} \tag{7}$$

$$= \left(\frac{L}{W}\right)^2$$

where L and W are to be determined. Replacing L by W+M in (7), we have

$$4x^2(x^2-1)W+4x^4(x^2-1)^3 = 2WM+M^2. \tag{8}$$

Letting $M = 2R(x^2-1)$ in (8) we now obtain

$$x^2W+x^4(x^2-1)^2 = WR+(x^2-1)R^2, \quad x \neq \pm 1$$

or

$$(x^2-1)(x^4(x^2-1)-R^2) = W(R-x^2). \tag{9}$$

Replacing R by $2x^2-1$ and W by $x^4(x^2-1)-R^2$ in (9) yields

$$W = x^6-5x^4+4x^2-1. \tag{10}$$

Since $R = 2x^2-1$, $M = 2R(x^2-1)$ and $L = W+M$, we see that

$$L = x^6-x^4-2x^2+1 \tag{11}$$

so that CE+1 is now a perfect square.

To make AE+1 a perfect square, we let

$$EA+1 = \frac{4DWA+D^3A}{4W^2}+1 = \left(\frac{Q}{W}\right)^2 \tag{12}$$

where Q is to be determined.

With $Q = W+J$ and $J = (x^2-x)5$, it can be shown that

$$Q = x^6-4x^4-2x^3+4x^2+x-1 \tag{13}$$

by following an argument similar to the one used to find W and L in (10) and (11). Likewise, it can be shown that with $(BE+1) = (N/W)^2$, we have

$$N = x^6-4x^4+2x^3+4x^2-x-1. \tag{14}$$

Hence, it has now been shown that when (5) is added to (1) - (4) with $W = x^6-5x^4+4x^2-1$, $x \neq \pm 1$ then there are 5 numbers such that the product of any two of the numbers plus 1 is a perfect square.

Note that E is not an integer which is in agreement with the result by Davenport and Baker.

Without going through the algebraic details, which are similar to those above, we merely mention that the set

$$\emptyset = \left\{\frac{NY^2-1}{4Y}, \frac{9NY^2-1}{Y}, \frac{25NY^2-9}{16Y}, \frac{49NY^2-1}{16Y}\right\}$$

has the property that the product of any two members when increased by N is a perfect square. In particular when $N = 1$ and $Y = 1$, we have

$$\emptyset = \{0, 8, 1, 3\}.$$

Similarly, the set

$$\psi = \left\{\frac{NY^2+1}{4Y}, \frac{9Y^2N+1}{Y}, \frac{25NY^2+9}{16Y}, \frac{49NY^2+1}{16Y}\right\}$$

has the property that the product of any two members in the set minus N is a perfect square. Here, when $Y = N = 1$, we see that

$$\psi = \left\{ \frac{1}{2}, 10, \frac{17}{8}, \frac{25}{8} \right\} .$$

If we consider the Fibonacci numbers defined recursively by

$$F_n = F_{n-1} + F_{n-2}, \ n \geq 3$$

with $F_1 = F_2 = 1$, we can also give an algebraic argument to show that the numbers in

$$\emptyset_1 = \left\{ \frac{F_{12u} - F_{12k}}{4}, 9F_{12u} - F_{12k}, \frac{25F_{12u} - 9F_{12k}}{16}, \frac{49F_{12u} - F_{12k}}{16} \right\}$$

have the property the product of any two increased by $n = F_{12u}F_{12k}$ is a square integer while the numbers in

$$\psi_1 = \left\{ \frac{F_{12u} + F_{12k}}{4}, 9F_{12u} + F_{12k}, \frac{25F_{12u} + 9F_{12k}}{16}, \frac{49F_{12u} + F_{12k}}{16} \right\}$$

have the property that the product of any two decreased by $n = F_{12u}F_{12k}$ is a square integer. In both cases, we require that $u \geq 1$, $k \geq 1$ and $u > k$. If we let $u = 2$ and $k = 1$ then $n = F_{12}F_{24} = 6,676,992$ and the sets are

$$\emptyset_1 = \{11556, 417168, 72369, 142002\}$$

and

$$\psi_1 = \{11628, 417456, 72531, 142011\}.$$

REFERENCES

[1] Arkin, J., Hoggatt, Jr., V. E. and Straus, E. G. "On Euler's Solution to a Problem of Diophantus." *The Fibonacci Quarterly, Vol. 17* (1979): pp 333-339.

[2] Arkin, J., Hoggatt, Jr., V. E. and Straus, E. G. "On Euler's Solution to a Problem of Diophantus-II." *The Fibonacci Quarterly, Vol. 18* (1980): pp 170-176.

[3] Davenport, H. and Baker, A. *Quarterly Journal of MATH, Vol. 20* (1969): pp 129-137.

[4] Diophantus, "Diophante d' Alexandrie" (ed. P. ver Eecke), (1959): pp 136-137.

[5] Fermat, P. "Observations sur Diophante." *(Observations Domini Petri de Fermat), Oe uvres de Fermat* (ed. P. Tannery and C. Henry), *Vol. 1 (MDCCCXCI)*, p 393.

[6] Hoggatt, Jr., V. E. and Bergum, G. E. "A Problem of Fermat and the Fibonacci Sequence." *The Fibonacci Quarterly, Vol. 15, No. 4* (1977): pp 323-330.

[7] Jones, W. J. "A Second Variation on a Problem of Diophantus and Davenport." *The Fibonacci Quarterly, Vol. 16, No. 2* (1978): pp 155-165.

[8] Jones, W. J. "A Variation on a Problem of Davenport and Diophantus." *Quarterly Journal of Math., Vol. 27* (1976): pp 349-353.

[9] Long, C. and Bergum, G. E. "On a Problem of Diophantus." *This publication.* pp 183-191.

Calvin Long and Gerald Bergum

ON A PROBLEM OF DIOPHANTUS

INTRODUCTION

Long ago Diophantus of Alexandria [4] noted that the numbers 1/16, 33/16, 68/16, and 105/16 all have the property that the product of any two increased by 1 is the square of a rational number. Much later, Fermat [5] notes that the product of any two of 1, 3, 8, and 120 increased by 1 is the square of an integer. In 1969, Davenport and Baker [3] showed that if the integers 1, 3, 8 and c have this property then c must be 120. From this it follows that there does not exist an integer d different from 1, 3, 5, and 120 such that the five numbers 1, 3, 8, 120, and d have the same property. In 1977, Hoggatt and Bergum [7] noted that $1 = F_2$, $3 = F_4$, $8 = F_6$ and $120 = 4 \cdot 2 \cdot 3 \cdot 5 = 4F_1F_3F_5$ where F_n is the nth Fibonacci number, and were led to guess that the numbers F_{2n}, F_{2n+2}, F_{2n+4}, and $4F_{2n+1}F_{2n+2}F_{2n+3}$ possessed this same property for every $n \geq 1$. Moreover, since 1, 2, and 5 have the property that the product of any two <u>decreased</u> by 1 is the square of an integer and $1 = F_1$, $2 = F_3$, and $5 = F_5$, they guessed that there must exist an integer y such that the product of any two of 1, 2, 5, and y decreased by 1 must be a perfect square. More generally, they guessed that for every $n \geq 0$ there must exist an integer y_n such that the product of any two of F_{2n+1}, F_{2n+3}, F_{2n+5}, and y_n decreased by 1 must be a perfect square. Their guesses were only partly correct; however, they were able to prove the following theorems.

<u>Theorem 1</u>: If F_n denotes the nth Fibonacci number and $x = 4F_{2n+1}F_{2n+2}F_{2n+3}$, then

$$F_{2n}F_{2n+2} + 1 = F_{2n+1}^2,$$
$$F_{2n}F_{2n+4} + 1 = F_{2n+2}^2,$$
$$F_{2n+2}F_{2n+4} + 1 = F_{2n+3}^2,$$
$$F_{2n}x + 1 = (2F_{2n+1}F_{2n+2}-1)^2,$$
$$F_{2n+2}x + 1 = (2F_{2n+1}F_{2n+3}-1)^2,$$
$$F_{2n+4}x + 1 = (2F_{2n+2}F_{2n+3}+1)^2.$$

A. N. Philippou et al. (eds.), Applications of Fibonacci Numbers, 183–191.
© 1988 by Kluwer Academic Publishers.

<u>Theorem 2</u>: If F_n denotes the nth Fibonacci number and $y = 4F_{2n+1}F_{2n+2}F_{2n+3}$, then

$$F_{2n+1}F_{2n+3} - 1 = F_{2n+2}^2,$$
$$F_{2n+1}F_{2n+5} - 1 = F_{2n+3}^2,$$
$$F_{2n+3}F_{2n+5} - 1 = F_{2n+4}^2,$$
$$F_{2n+1}y + 1 = (2F_{2n+2}F_{2n+3}+1)^2,$$
$$F_{2n+3}y + 1 = (2F_{2n+2}F_{2n+4}+1)^2,$$
$$F_{2n+5}y + 1 = (2F_{2n+3}F_{2n+4}-1)^2.$$

They also conjectured but could not prove that the x and y of the preceding Theorems were unique.

Of course, it is only natural to ask if similar results hold for the Lucas numbers

$$1, 3, 4, 7, 11, 18, 29, 47, \ldots$$

The answer, as we show here, is a qualified yes.

2. THE LUCAS CASE

For a second order recurrence $\{f_n\}$, let $f_1f_3 - f_2^2 = c$ be the <u>characteristic</u> of the sequence. It turns out that the characteristic plays a key role in these results. Thus, for the Fibonacci sequence, $c = F_1F_3 - F_2^2 = 1 \cdot 2 - 1 = 1$ and 1 plays a key role in Theorems 1 and 2. For the Lucas sequence $c = L_1L_3 - L_2^2 = 1 \cdot 4 - 9 = -5$ and -5 plays a similar role in the following.

<u>Theorem 3</u>: If L_n denotes the nth Lucas number, then

$$L_{2n}L_{2n+2} - 5 = L_{2n+1}^2,$$
$$L_{2n}L_{2n+4} - 5 = L_{2n+2}^2,$$
$$L_{2n+2}L_{2n+4} - 5 = L_{2n+3}^2.$$

<u>Proof</u>: The results are an immediate consequence of Binet's formulas.

Theorem 4: If L_n denotes the nth Lucas number, then

$$L_{2n+1}L_{2n+3} + 5 = L_{2n+2}^2,$$
$$L_{2n+1}L_{2n+5} + 5 = L_{2n+3}^2,$$
$$L_{2n+3}L_{2n+5} + 5 = L_{2n+4}^2.$$

Proof: Again the results immediately follow from Binet's formulas.

Of course, the analogy with the first three equations of each of Theorem 1 and Theorem 2 is clear and precise. One simply replaces Fibonacci numbers by Lucas numbers and the Fibonacci characteristic $c = 1$ by the Lucas characteristic $c = -5$ in each case. However, the strict analogy with the last three equations in each Theorem breaks down. Indeed, the following Theorems show that the analogy to the last half of Theorems 1 and 2 do not hold for Lucas numbers.

Theorem 5: Let L_n denote the nth Lucas number. Then there does not exist an integer x and integers r, s, and t such that

$$\text{(i)} \quad L_{2n}x + 5 = r^2,$$
$$\text{(ii)} \quad L_{2n+2}x + 5 = s^2, \tag{1}$$
$$\text{(iii)} \quad L_{2n+4}x + 5 = t^2.$$

Proof: Eliminating x between 1-(i), (ii), and (iii) in pairs, we obtain

$$\text{(i)} \quad 5L_{2n+2} - 5L_{2n} = L_{2n+2}r^2 - L_{2n}s^2,$$
$$\text{(ii)} \quad 5L_{2n+4} - 5L_{2n} = L_{2n+4}r^2 - L_{2n}t^2, \tag{2}$$
$$\text{(iii)} \quad 5L_{2n+4} - 5L_{2n+2} = L_{2n+4}s^2 - L_{2n+2}t^2.$$

We now show that (2) leads to a contradiction so that no such x, r, s, and t exist. Our argument is based on the fact that the Lucas sequence is periodic modulo 8 with period of length 12 and residues as indicated in the following table.

n mod 12	1	2	3	4	5	6	7	8	9	10	11	12
L_n mod 8	1	3	4	7	3	2	5	7	4	3	7	2

Case 1: n = 3k, k even.

Since L_{6k} is even, it follows from 1-(i) that r is odd and hence $r^2 \equiv$ 1(mod 8). Therefore, 2-(i) yields

$$5L_{6k+2} - 5L_{6k} = L_{6k+2}r^2 - L_{6k}s^2,$$
$$5 \cdot 3 - 5 \cdot 2 \equiv 3 \cdot 1 - 2s^2 \text{ (mod 8)},$$
$$2 \equiv -2s^2 \text{ (mod 8)},$$

or finally

$$s^2 \equiv -1 \text{ (mod 4)}.$$

But this is a contradiction so that this case is impossible.

Because the arguments in each of the other five cases are exactly analogous to Case 1 we omit the details and the proof is complete.

Theorem 6: Let L_n denote the nth Lucas number. Then there does not exist an integer y and integers r, s, and t such that

$$\text{(i)} \quad L_{2n+1}y + 5 = r^2,$$
$$\text{(ii)} \quad L_{2n+3}y + 5 = s^2, \tag{3}$$
$$\text{(iii)} \quad L_{2n+5}y + 5 = t^2,$$

Proof: Eliminating y between 3-(i), (ii), and (iii) in pairs, we obtain

$$\text{(i)} \quad 5L_{2n+3} - 5L_{2n+1} = L_{2n+3}r^2 - L_{2n+1}s^2,$$
$$\text{(ii)} \quad 5L_{2n+5} - 5L_{2n+1} = L_{2n+5}r^2 - L_{2n+1}t^2,$$
$$\text{(iii)} \quad 5L_{2n+5} - 5L_{2n+3} = L_{2n+5}s^2 - L_{2n+3}t^2.$$

Once again we show that these lead to a contradiction. Here, however, it is necessary to argue both modulo 8 and modulo 4. Modulo 4, the Lucas sequence is periodic of period 6 with residues as indicated in the following table.

n mod 6	1	2	3	4	5	6
L_n mod 8	1	3	0	3	3	2

We have the same six cares as before, but since the arguments are all the same we prove only

Case 1: n = 3k, k even.

Since L_{6k+3} is even, it follows from 3-(ii) that s is odd. Hence, 4-(iii) yields

$$5L_{6k+5} - 5L_{6k+3} = L_{6k+5}s^2 - L_{6k+3}t^2,$$
$$5\cdot3 - 5\cdot4 \equiv 3\cdot1 - 4t^2 \ (\text{mod } 8),$$
$$-8 \equiv -4t^2 \ (\text{mod } 8)$$

so that t is even. But then 3-(iii) implies that

$$L_{6k+5}x + 5 \equiv 0 \ (\text{mod } 4),$$
$$3x \equiv -1 \ (\text{mod } 4),$$
$$x \equiv 1 \ (\text{mod } 4)$$

while 3-(i) gives

$$L_{6k+1}\cdot1 + 5 \equiv r^2 \ (\text{mod } 4),$$
$$1\cdot1 + 1 \equiv r^2 \ (\text{mod } 4),$$
$$2 \equiv r^2 \ (\text{mod } 4)$$

which is a contradiction. The other five cases may be treated in exactly the same way to complete the proof.

3. A MORE GENERAL CASE

Now consider the sequence $\{f_n\}$ defined by

$$f_1 = a, \quad f_2 = b$$
$$f_{n+2} = f_{n+1} + f_n \qquad \forall_n \geq 1 \tag{5}$$

We ask for what integer values of a and b do there exist analogs of Theorem 1 and Theorem 2.

It is easilty shown that

$$f_{n+2} = bF_{n+1} + aF_n, \tag{6}$$

where F_n denotes the nth Fibonacci number as above, and that

$$f_n = \frac{\alpha^{n-2}(a+b\alpha) - \beta^{n-2}(a+b\beta)}{\sqrt{5}} \tag{7}$$

where $\alpha = (1 + \sqrt{5})/2$ and $\beta = (1 - \sqrt{5})/2$. Also, it is easy to show that the characteristic is

$$c = a^2 + ab - b^2 \tag{8}$$

We then have the following theorems which are analogous to Theorem 1 and Theorem 2.

<u>Theorem 7</u>: If $\{f_n\}$ is as defined in (5) and $x = 4f_{2n+1}f_{2n+2}f_{2n+3}$, then

$$f_{2n}f_{2n+2} + c = f_{2n+1}^2,$$
$$f_{2n}f_{2n+4} + c = f_{2n+2}^2,$$
$$f_{2n+2}f_{2n+4} + c = f_{2n+3}^2,$$
$$f_{2n}x + c^2 = (2f_{2n+1}f_{2n+2} - c)^2,$$
$$f_{2n+2}x + c^2 = (2f_{2n+1}f_{2n+3} - c)^2,$$
$$f_{2n+4}x + c^2 = (2f_{2n+2}f_{2n+3} + c)^2.$$

<u>Theorem 8</u>: If $\{f_n\}$ is as defined in (5) and $y = 4f_{2n+2}f_{2n+3}f_{2n+4}$, then

$$f_{2n+1}f_{2n+3} - c = f_{2n+2}^2,$$
$$f_{2n+1}f_{2n+5} - c = f_{2n+3}^2,$$

$$f_{2n+3}f_{2n+5} - c = f_{2n+4}^2,$$
$$f_{2n+1}y + c^2 = (2f_{2n+2}f_{2n+3} + c)^2,$$
$$f_{2n+3}y + c^2 = (2f_{2n+2}f_{2n+4} + c)^2,$$
$$f_{2n+5}y + c^2 = (2f_{2n+3}f_{2n+4} - c)^2.$$

The proofs of these two theorems follow directly from (7) and (8) and the fact that

$$c = (a+b\alpha)(a+b\beta).$$

The details are omitted.

As should be the case, when $a = b = 1$, we have $f_n = F_n$ $\forall n$ with $c = 1$ so that Theorem 7 gives the equations of Theorem 1 and Theorem 8 gives the equations of Theorem 2.

Letting $a = 1$ and $b = 3$, we have $f_n = L_n$ $\forall n$ and $c = a^2 + ab - b^2 = 1 + 3 - 9 = -5$. In this case, Theorem 7 yields the following result.

Corollary 9: If L_n denotes the nth Lucas number and $x = 4L_{2n+1}L_{2n+2}L_{2n+3}$, then

$$L_{2n}L_{2n+2} - 5 = L_{2n+1}^2,$$
$$L_{2n}L_{2n+4} - 5 = L_{2n+2}^2,$$
$$L_{2n+2}L_{2n+4} - 5 = L_{2n+3}^2,$$
$$L_{2n}x + 25 = (2L_{2n+1}L_{2n+2} + 5)^2,$$
$$L_{2n+2}x + 25 = (2L_{2n+1}L_{2n+3} + 5)^2,$$
$$L_{2n+4}x + 25 = (2L_{2n+2}L_{2n+3} - 5)^2.$$

Similarly, Theorem 8 yields the result.

Corollary 10: If L_n denotes the nth Lucas number and $y = 4L_{2n+2}L_{2n+3}L_{2n+4}$, then

$$L_{2n+1}L_{2n+3} + 5 = L_{2n+2}^2,$$
$$L_{2n+1}L_{2n+5} + 5 = L_{2n+3}^2,$$
$$L_{2n+3}L_{2n+5} + 5 = L_{2n+4}^2,$$

$$L_{2n+1}y + 25 = (2L_{2n+2}L_{2n+3} - 5)^2,$$
$$L_{2n+3}y + 25 = (2L_{2n+2}L_{2n+4} - 5)^2,$$
$$L_{2n+6}y + 25 = (2L_{2n+3}L_{2n+4} + 5)^2.$$

Of course, Theorem 6 and Theorem 7 are analogous to Theorem 1 and Theorem 2. But the analogy is not exact since, in each case, we add c^2 rather than c on the left hand side of the last three equations. To remedy this situation it is necessary that $c^2 = c$ which implies that $c = 0$ or $c = 1$. The first of these conditions requires that $a = (-b \pm b\sqrt{5})/2$ and hence that $a = b = 0$ if a and b are to be integers. This gives the null sequence and is therefore of no interest. The second condition gives

$$a = \frac{-b \pm \sqrt{5b^2 + 4}}{2}.$$

Here a will be an integer if and only if b is an integer such that $m = \sqrt{5b^2 + 4}$ is also an integer; that is, if and only if

$$m^2 - 5b^2 = 4.$$

But by [11], this is so if and only if $b = F_{2r}$ for some integer r and

$$a = \frac{-F_{2r} \pm \sqrt{5F_{2r}^2 + 4}}{2} = \frac{-F_{2r} \pm L_{2r}}{2}$$

where F_n and L_n denote the nth Fibonacci and Lucas numbers as above. If we take the plus sign $a = F_{2r-1}$ and we obtain the sequence $\{f_n\}$ where $f_n = F_{2r+n-2}$ for all n. Similarly, if we take the minus sign $a = -F_{2r-1}$ and we obtain the sequence $\{f_n\}$ where $f_n = (-1)^n F_{2r-n+2}$ for all n. Thus, since $F_{-n} = (-1)^{n-1}F_n$, it turns out that essentially the only sequence that has properties exactly analogous to those expressed in Theorem 1 and Theorem 2 is the Fibonacci sequence.

Of course, one should go on to consider the more general equation defined by $g_1 = r$, $g_2 = s$, $g_{n+2} = ag_{n+1} + bg_n$ for all n. Clearly theorems can be obtained, but we desist at this point since we understand that Alwyn Horadam pursues these results in a paper [8].

REFERENCES

[1] Arkin, J. and Bergum, G. "More on the Problem of Diophantus." *This publication*, pp 177-181.

[2] Berzenyi, G. *B-369, The Fibonacci Quarterly, Vol. 16, No. 6* (1978): p 565.

[3] Davenport, H. and Baker, A. *Quarterly Journal of MATH, Vol. 20* (1969): pp 129-137.

[4] Diophantus, "Diophante d'Alexandrie." (ed. P. ver Eecke), (1959): pp 136-137.

[5] Fermat, P. "Observations sur Diophante." *(Observations Domini Petri de Fermat), Oe uvres de Fermat (ed. P. Tannery and C. Henry), Vol 1. (MDCCCXCI)*, p 393.

[6] Gupta, H. and Singh, K. "On k-Triad Sequences." *Internat. J. Math. & Math. Sci., Vol. 5, No. 4* (1985): pp 799-804

[7] Hoggatt, Jr. V. E. and Bergum, G. E. "A Problem of Fermat and the Fibonacci Sequence." *The Fibonacci Quarterly, Vol. 15, No. 4* (1977): pp 323-330.

[8] Horadam, A. F. "Generalization of a Result of Morgado." *to appear.*

[9] Jones, B. W. "A Second Variation on a Problem of Diophantus and Davenport." *The Fibonacci Quarterly, Vol. 16, No. 2* (1978): pp 155-165.

[10] Jones, B. W. "A Variation on a Problem of Davenport and Diophantus." *Quarterly Journal of Math., Vol. 27* (1976): pp 349-353.

[11] Long, C. T. and Jordan, J. H. "A Limited Arithmetic on Simple Continued Fractions." *The Fibonacci Quarterly, Vol. 8, No. 2* (1970): pp 135-157.

Marjorie Bicknell-Johnson and Gerald E. Bergum

THE GENERALIZED FIBONACCI NUMBERS
$\{C_n\}$, $C_n = C_{n-1} + C_{n-2} + K$

INTRODUCTION

There are many ways to generalize the well-known Fibonacci sequence $\{F_n\}$, $F_n = F_{n-1} + F_{n-2}$, $F_1 = 1$, $F_2 = 1$. In a personal letter dated December 18, 1985, Frank Harary asked one of the authors if they had ever encountered $C_n = C_{n-1} + C_{n-2} + 1$, which was used by Harary in connection with something he was counting involving Boolean Algebras. In fact, in Harary's research it was noticed that the value of one could be replaced by any integer k.

Furthermore, one of the authors noticed recently that the integers generated by the sequence, with $C_0 = C_1 = 1$, are related to the number of nodes found in a Fibonacci tree, [1], [2].

These observations culminated into the output of this paper and some questions with unknown answers.

1. SPECIAL SEQUENCES

If we generalize the Fibonacci sequence $\{F_n\} = \{1, 1, 2, 3, 5, 8, \ldots\}$ by adding one at each step of the recursion, we form a new sequence $\{c_n\} = \{1, 1, 3, 5, 9, 15, 25, 41, \ldots\}$ where

$$c_{n+1} = c_n + c_{n-1} + 1, \ c_1 = 1, \ c_2 = 1. \tag{1.1}$$

It is not difficult to show that

$$c_n = 2F_n - 1, \tag{1.2}$$

A. N. Philippou et al. (eds.), Applications of Fibonacci Numbers, 193–205.

and to extend the subscripts so that n is any integer.

However, more fruitful special sequences can be formed by changing the beginning values in (1.1). If we write $\{c_n^*\} = \{1, 2, 4, 7, 12, 20, 33, 54, 88, \ldots\}$ where $c_1^* = 1$ and $c_2^* = 2$ in the recursion of (1.1), we find that

$$c_n^* = F_{n+2} - 1. \tag{1.3}$$

Changing the initial values to zero, results in $\{c_n'\} = \{0, 0, 1, 2, 4, 7, 12, 20, \ldots\}$ which has the same terms occurring as in $\{c_n^*\}$ but

$$c_n' = F_n - 1. \tag{1.4}$$

Actually, the sequences $\{c_n\}$, $\{c_n^*\}$, and $\{c_n'\}$ are all special cases of the more general sequence $\{C_n(a,b)\}$, where

$$C_{n+1}(a,b) = C_n(a,b) + C_{n-1}(a,b) + 1, \; C_1(a,b) = a, \; C_2(a,b) = b. \tag{1.5}$$

The first few terms of $\{C_n(a,b)\}$ are

$$a, \; b, \; a + b + 1, \; a + 2b + 2, \; 2a + 3b + 4, \; 3a + 5b + 7, \; 5a + 8b + 12, \ldots$$

$$C_n(a,b) = aF_{n-2} + bF_{n-1} + F_n - 1, \quad n \ge 1, \tag{1.6}$$

and by standard techniques the generating function $g(a,b;x)$ for $\{C_n(a,b)\}$ is

$$g(a,b;x) = \sum_{n=1}^{\infty} C_n(a,b)x^{n-1} = \frac{a + (b - 2a)x + (a - b + 1)x^2}{1 - 2x + x^3} \tag{1.7}$$

Since a and b are arbitrary, one could investigate many possibilities. We note three cases that we found to be most interesting.

When $a = b = 1$, $\{C_n(1,1)\}$ is generated by

$$g(1,1;x) = \frac{1 - x + x^2}{1 - 2x + x^3},$$

and, as in (1.2),

$$C_n(1,1) = 2F_n - 1. \qquad (1.8)$$

If $a = 1$ and $b = 2$, then

$$g(1,2;x) = \frac{1}{1 - 2x + x^3}$$

and, as in (1.3),

$$C_n(1,2) = F_{n-2} + 2F_{n-1} + F_n - 1 = F_{n+2} - 1. \qquad (1.9)$$

Also, note that for $a = b = 0$, $C_n(0,0) = c'_n$ from (1.4), and

$$g(0,0;x) = \frac{x^2}{1 - 2x + x^3},$$

with

$$C_n(0,0) = F_n - 1. \qquad (1.10)$$

In passing, we note that, since $\displaystyle\lim_{n \to \infty} \frac{F_{n+k}}{F_n} = \left[\frac{1 + \sqrt{5}}{2}\right]_k$,

$$\lim_{n \to \infty} \frac{C_{n+1}(a,b)}{C_n(a,b)} = \frac{1 + \sqrt{5}}{2}. \qquad (1.11)$$

One can easily extend $\{C_n(a,b)\}$ to negative subscripts, and show that (1.8) holds for all integers n. Then, since $C_0(a,b) = b - a - 1$ and $C_{-1}(a,b) = 2a - b$, we can rewrite the generating function for $\{C_n(a,b)\}$ from (1.9) as

$$g(a,b;x) = \sum_{n=1}^{\infty} C_n(a,b)x^{n-1} = \frac{a - C_{-1}(a,b)x - C_0(a,b)x^2}{1 - 2x + x^3}$$

Furthermore, we note that $\{C_n(1,2)\}$ and $\{C_n(0,0)\}$ are important special cases, since (1.6) can be rewritten as

$$C_n(a,b) = aF_{n-2} + bF_{n-1} + C_{n-2}(1,2) \qquad (1.12)$$

or

$$C_n(a,b) = aF_{n-2} + bF_{n-1} + C_n(0,0). \tag{1.13}$$

In fact, (1.12) and (1.13) lead to multitudinous special cases:

$$C_n(1,2) = F_{n+1} + C_{n-2}(1,2) \tag{1.14}$$
$$C_n(0,0) = C_{n-2}(1,2) \tag{1.15}$$
$$C_n(0,b) = bF_{n-1} + C_n(0,0) \tag{1.16}$$
$$C_n(a,0) = aF_{n-2} + C_n(0,0) \tag{1.17}$$
$$C_n(a,a) = aF_n + C_n(0,0) \tag{1.18}$$
$$C_n(a,2a) = aF_{n+1} + C_n(0,0) \tag{1.19}$$
$$C_n(-a,a) = aF_{n-3} + C_n(0,0) \tag{1.20}$$
$$C_n(F_m,F_{m+1}) = F_{m+n-1} + C_n(0,0) \tag{1.21}$$

Since the Lucas sequence $\{L_n\}$ is given by

$$L_{n+1} = L_n + L_{n-1}, \; L_1 = 1, \; L_2 = 3,$$

we have the following additional special cases:

$$C_n(2,1) = L_{n-1} + C_n(0,0) \tag{1.22}$$
$$C_n(3,1) = 2L_{n-1} - 1 \tag{1.23}$$
$$C_n(1,3) = L_n + C_n(0,0) = 2F_{n+1} - 1 \tag{1.24}$$
$$C_n(2,3) = L_{n+1} - 1 \tag{1.25}$$
$$C_n(3,4) = L_{n+1} + C_n(0,0) \tag{1.26}$$
$$C_n(2a,a) = aL_{n-1} + C_n(0,0) \tag{1.27}$$
$$C_n(a,3a) = aL_n + C_n(0,0) \tag{1.28}$$
$$C_n(-a,2a) = aL_{n-2} + C_n(0,0) \tag{1.29}$$
$$C_n(L_m, L_{m+1}) = L_{m+n-1} + C_n(0,0) \tag{1.30}$$

Returning to (1.6), we can easily derive

$$C_n(a+1,b+1) = F_n + C_n(a,b) \tag{1.31}$$

and

$$C_n(a+m,b+m) = mF_n + C_n(a,b).$$ (1.32)

Using (1.14), we can change (1.12) to

$$C_n(a,b) = aF_{n-2} + bF_{n-1} + [C_n(1,2) - F_{n+1}]$$

so that

$$C_n(a,b) = (a - 1)F_{n-2} + (b - 2)F_{n-1} + C_n(1,2)$$ (1.33)

leading to

$$C_n(1,b) = (b - 2)F_{n-1} + C_n(1,2)$$ (1.34)

$$C_n(a,2) = (a - 1)F_{n-2} + C_n(1,2)$$ (1.35)

$$C_n(a,2a) = (a - 1)F_{n+1} + C_n(1,2)$$ (1.36)

$$C_n(3,1) = F_{n-4} + C_n(1,2)$$ (1.37)

$$C_n(3,4) = 2F_n + C_n(1,2)$$ (1.38)

$$C_n(a,a+1) = (a - 1)F_n + C_n(1,2)$$ (1.39)

No attempt has been made to be exhaustive, but it is time to turn to some special summations.

Since

$$\sum_{k=1}^{n} F_k = F_{n+2} - 1,$$ (1.40)

$C_n(1,2)$ is the sum of the first n Fibonacci numbers. By using (1.6) and (1.40),

$$\sum_{k=1}^{n} C_k(a,b) = a \sum_{k=1}^{n} F_{k-2} + b \sum_{k=1}^{n} F_{k-1} + \sum_{k=1}^{n} F_k - n$$

$$= a \sum_{k=-1}^{n-2} F_k + b \sum_{k=1}^{n-1} + \sum_{k=1}^{n} F_k - n$$

$$= a(F_{-1} + F_0 + F_n - 1) + b(F_{n+1} - 1) + (F_{n+2} - 1) - n$$

$$= (aF_n + bF_{n+1} + F_{n+2} - 1) - b - n$$

$$= C_{n+2}(a,b) - b - n.$$

Thus,

$$\sum_{k=1}^{n} C_k(a,b) = C_{n+2}(a,b) - (b + n) \tag{1.41}$$

with special cases

$$\sum_{k=1}^{n} C_k(1,2) = C_{n+2}(1,2) - (n + 2) = F_{n+4} - (n + 3) \tag{1.42}$$

$$\sum_{k=1}^{n} C_k(1,1) = C_{n+2}(1,1) - (n + 1) = 2F_{n+2} - (n + 2) \tag{1.43}$$

$$\sum_{k=1}^{n} C_k(0,0) = C_{n+2}(0,0) - n = F_{n+2} - (n + 1) \tag{1.44}$$

Since

$$\sum_{k=1}^{n} kF_{k+p} = (n + 1)F_{n+2+p} - F_{n+4+p} + F_{3+p}, \tag{1.45}$$

we can use (1.6) to derive

$$\sum_{k=1}^{n} kC_k(a,b) = a \sum_{k=1}^{n} kF_{k-2} + b \sum_{k=1}^{n} kF_{k-1} + \sum_{k=1}^{n} kF_k - \sum_{k=1}^{n} k$$

$$= a[(n + 1)F_n - F_{n+2} + F_1] + b[(n+1)F_{n+1} - F_{n+3} + F_2]$$
$$+ [(n + 1)F_{n+2} - F_{n+4} + F_3] - n(n + 1)/2$$
$$= (n + 1)(aF_n + bF_{n+1} + F_{n+2})$$
$$- (aF_{n+2} + bF_{n+3} + F_{n+4}) + (aF_1 + bF_2 + F_3) - n(n + 1)/2$$
$$= (n + 1)[C_{n+2}(a,b) + 1] - [C_{n+4}(a,b) + 1]$$
$$+ [C_3(a,b) + 1] - n(n+1)/2$$
$$= (n + 1)C_{n+2}(a,b) - C_{n+4}(a,b) + C_3(a,b) - (n - 2)(n + 1)/2$$

so that

$$\sum_{k=1}^{n} kC_k(a,b) = (n + 1)C_{n+2}(a,b) - C_{n+4}(a,b) + C_3(a,b) \tag{1.46}$$

$$- \frac{(n - 2)(n + 1)}{2} .$$

Looking to

$$F_n^2 - F_{n+1}F_{n-1} = (-1)^{n+1}, \tag{1.47}$$

with patient algebra, one can derive

$$C_n^2(a,b) - C_{n-1}(a,b)C_{n+1}(a,b) = (a^2 - b^2 + 3a + ab - b + 1)(-1)^{n+1}$$
$$+ C_{n-3}(a,b) + 1 \tag{1.48}$$

from (1.6).

Suppose that we put $\{C_n(a,b)\}$ into a 3x3 determinant as below, where we let $C_n = C_n(a,b)$. Using (1.5), we subtract the sum of the first and second columns from the third column, followed by the same sequence for the rows. Then, add the third column to the first and second columns.

$$D_n = \begin{vmatrix} C_n & C_{n+1} & C_{n+2} \\ C_{n+1} & C_{n+2} & C_{n+3} \\ C_{n+2} & C_{n+3} & C_{n+4} \end{vmatrix} = \begin{vmatrix} C_n & C_{n+1} & 1 \\ C_{n+1} & C_{n+2} & 1 \\ C_{n+2} & C_{n+3} & 1 \end{vmatrix} = \begin{vmatrix} C_n & C_{n+1} & 1 \\ C_{n+1} & C_{n+2} & 1 \\ 1 & 1 & -1 \end{vmatrix}$$

$$= \begin{vmatrix} C_n + 1 & C_{n+1} + 1 & 1 \\ C_{n+1} + 1 & C_{n+2} + 1 & 1 \\ 0 & 0 & -1 \end{vmatrix}$$

$$= (-1)[(C_n + 1)(C_{n+2} + 1) - (C_{n+1} + 1)^2]$$
$$= (C_{n+1}^2 + 2C_{n+1} + 1) - (C_n C_{n+2} + C_n + C_{n+2} + 1)$$
$$= C_{n+1}^2 - C_n C_{n+2} + (C_{n+1} - C_n) - (C_{n+2} - C_{n+1}).$$

Returning to (1.5), notice that

$$C_{n+1}(a,b) - C_n(a,b) = C_{n-1}(a,b) + 1 \tag{1.49}$$

and continue to derive D_n.

$$D_n = (C_{n+1}^2 - C_n C_{n+2}) + (C_{n-1} + 1) - (C_n - 1)$$
$$= (C_{n+1}^2 - C_n C_{n+2}) - C_{n-2} - 1$$
$$[(a^2 - b^2 + 3a + ab - b + 1)(-1)^n + C_{n-2} + 1] - C_{n-2} - 1$$

by (1.48). Hence, we have

$$D_n = (a^2 - b^2 + 3a + ab - b + 1)(-1)^n. \qquad (1.50)$$

Note that, if we use $\{C_n(1,1)\}$ to form D_n, then $D_n = 4(-1)^n$, while $D_n = (-1)^n$ for $\{C_n(0,0)\}$ or for $\{C_n(1,2)\}$.

2. THE GENERALIZED SEQUENCE $C_n(a,b,k)$

When we let $C_n(a,b,k)$ be defined as

$$C_n(a,b,k) = C_{n-1}(a,b,k) + C_{n-2}(a,b,k) + k, \; C_1 = a, \; C_2 = b, \qquad (2.1)$$

the first few terms

$$a, \; b, \; a + b + k, \; a + 2b + 2k, \; 2a + 3b + 4k, \; 3a + 5b + 7k,$$
$$5a + 8b + 12k, \; 8a + 13b + 20k, \ldots,$$

immediately suggest

$$C_n(a,b,k) = aF_{n-2} + bF_{n-1} + kC_n(0,0), \qquad (2.2)$$
$$C_n(a,b,k) = aF_{n-2} + bF_{n-1} + kF_n - k, \qquad (2.3)$$

where the subscripts can extend to any integer. Of course, $C_n(a,b,1) = C_n(a,b)$.

Applying (1.40) to (2.3), one can optain

$$\sum_{i=1}^{n} C_i(a,b,k) = aF_n + bF_{n+1} + kF_{n+2} - b - (n + 1)k \qquad (2.4)$$
$$= C_{n+2}(a,b,k) - b - kn.$$

Since $\displaystyle\lim_{n\to\infty} \frac{F_{n+k}}{F_n} = \left(\frac{1 + \sqrt{5}}{2}\right)^k$, elementary limit theorems lead to

$$\lim_{n\to\infty} \frac{C_{n+1}(a,b,k)}{C_n(a,b,k)} = \frac{1 + \sqrt{5}}{2} . \tag{2.5}$$

Using

$$C_n(0,0) = C_n(1,2) - F_{n+1}$$

from (1.14) and (1.15) in (2.2), one can convert $C_n(a,b,k)$ into a form in which $C_n(1,2)$ appears, writing

$$C_n(a,b,k) = (a - k)F_{n-2} + (b - 2k)F_{n-1} + kC_n(1,2). \tag{2.6}$$

Next, we use (2.2) and (2.6) to write some particular sequences $\{C_n(a,b,k)\}$ that have interesting general terms.

$$C_n(a,a,k) = aF_n + kC_n(0,0) \tag{2.7}$$
$$C_n(a,0,k) = aF_{n-2} + kC_n(0,0) \tag{2.8}$$
$$C_n(0,b,k) = bF_{n-1} + kC_n(0,0) \tag{2.9}$$
$$C_n(a,2a,k) = aF_{n+1} + kC_n(0,0) \tag{2.10}$$
$$C_n(F_m,F_{m+1},k) = F_{m+n-1} + kC_n(0,0) \tag{2.11}$$
$$C_n(L_m,L_{m+1},k) = L_{m+n-1} + kC_n(0,0) \tag{2.12}$$
$$C_n(a+m,b+m,k) = mF_m + C_n(a,b,k) \tag{2.13}$$
$$C_n(k,b,k) = (b - 2k)F_{n-1} + kC_n(1,2) \tag{2.14}$$
$$C_n(a,2k,k) = (a - k)F_{n-2} + kC_n(1,2) \tag{2.15}$$
$$C_n(b-k,b,k) = (b - 2k)F_n + kC_n(1,2) \tag{2.16}$$
$$C_n(a,2a,k) = (a - k)F_{n+1} + kC_n(1,2) \tag{2.17}$$
$$C_n(a,a-k,k) = aF_n + kC_{n-2}(0,0) \tag{2.18}$$

If $\{C_n(a,b,k)\}$ is put into a 3x3 determinant D_n, where we let $C_n = C_n(a,b,k)$, we can establish that

$$D_n = \begin{vmatrix} C_n & C_{n+1} & C_{n+2} \\ C_{n+1} & C_{n+2} & C_{n+3} \\ C_{n+2} & C_{n+3} & C_{n+4} \end{vmatrix} = (-1)^n k(a^2 - b^2 + k^2 + 3ak + ab - bk). \quad (2.19)$$

Since $C_{n-1} + C_n = C_{n+1} - k$,

$$D_n = \begin{vmatrix} C_n & C_{n+1} & C_{n+2} \\ C_{n+1} & C_{n+2} & C_{n+3} \\ C_{n+2} & C_{n+3} & C_{n+4} \end{vmatrix} = \begin{vmatrix} C_n & C_{n+1} & k \\ C_{n+1} & C_{n+2} & k \\ C_{n+2} & C_{n+3} & k \end{vmatrix}.$$

Since $C_n - C_{n+1} = -C_{n-1} - k$,

$$D_{n-1} = \begin{vmatrix} C_{n-1} & C_n & C_{n+1} \\ C_n & C_{n+1} & C_{n+2} \\ C_{n+1} & C_{n+2} & C_{n+3} \end{vmatrix} = \begin{vmatrix} -k & C_n & C_{n+1} \\ -k & C_{n+1} & C_{n+2} \\ -k & C_{n+2} & C_{n+3} \end{vmatrix},$$

and $D_{n-1} = (-1)D_n$. So, we only have to evaluate one of these determinants D_n.

$$D_0 = \begin{vmatrix} -a+b-k & a & b \\ a & b & a+b+k \\ b & a+b+k & a+2b+2k \end{vmatrix} = \begin{vmatrix} -a+b-k & a & k \\ a & b & k \\ k & k & -k \end{vmatrix}$$

$$= \begin{vmatrix} -a+b & b+k & 0 \\ a+k & b+k & 0 \\ k & k & -k \end{vmatrix}$$

$$= k(a^2 - b^2 + k^2 + 3ak + ab - bk).$$

Thus, since $D_{n-1} = (-1)D_n$, we have (2.19).

Since we also could write

$$D_{n-1} = \begin{vmatrix} C_{n-1} & C_n & k \\ C_n & C_{n+1} & k \\ k & k & -k \end{vmatrix} = \begin{vmatrix} C_{n-1} + k & C_n + k & 0 \\ C_n + k & C_{n-1} + k & 0 \\ k & k & -k \end{vmatrix} \quad (2.20)$$

$$= k[(C_n + k)^2 - (C_{n-1} + k)(C_{n+1} + k)],$$

we can derive another identity with a little patience. Observe that

$$(C_n + k)^2 - (C_{n-1} + k)(C_{n+1} + k) \quad (2.21)$$
$$= (C_n^2 - C_{n-1}C_{n+1}) - k(C_{n+1} + C_{n-1} - 2C_n)$$
$$= (C_n^2 - C_{n-1}C_{n+1}) - k(C_{n-3} + k)$$

since

$$F_{n+1} + F_{n-1} - 2F_n = F_{n-3} . \quad (2.22)$$

Thus, rearranging (2.21) and using (2.19) and (2.20),

$$C_n^2 - C_{n+1}C_{n-1} = (C_n + k)^2 - (C_{n+1} + k)(C_{n-1} + k) + k(C_{n-3} + k)$$
$$= (D_{n-1})/k + k(C_{n-3} + k)$$

establishing

$$C_n^2 - C_{n+1}C_{n-1} = (-1)^{n-1}(a^2 - b^2 + k^2 + 3ak + ab - bk) \quad (2.23)$$
$$+ kC_{n-3} + k^2$$

where $C_n = C_n(a,b,k)$. Compare with Equation (1.48).

Finally, we make a sum where we lean on Equation (1.45) several times, writing

$$\sum_{i=1}^{n} iC_i = \sum_{i=1}^{n} ia\, F_{i-2} + \sum_{i=1}^{n} ibF_{i-1} + \sum_{i=1}^{n} ikF_i - \sum_{i=1}^{n} ik$$

$$= a[(n + 1)F_n - F_{n+2} - 1] + b[(n + 1)F_{n+1} - F_{n+3} + 1]$$
$$+ k[(n + 1)F_{n+2} - F_{n+4} + 2] - kn(n + 1)/2$$
$$= (n + 1)(aF_n + bF_{n+1} + kF_{n+2} - k) + (n + 1)k$$
$$- (aF_{n+2} + bF_{n+3} + kF_{n+4} - k) - k$$

$$+ (a + b + k) + k - kn(n + 1)/2$$
$$= (n + 1)C_{n+2} - C_{n+4} + C_3 + (n+1)k - kn(n + 1)/2$$

finally yielding

$$\sum_{i=1}^{n} iC_i(a,b,k) = (n + 1)C_{n+2}(a,b,k) - C_{n+4}(a,b,k) + C_3(a,b,k) \quad (2.24)$$
$$- \frac{k(n - 2)(n + 1)}{2},$$

which has Equation (1.46) as a special case.

Similarly to (1.41), one could write

$$\sum_{i=1}^{n} C_i(a,b,k) = C_{n+2}(a,b,k) - b - kn. \tag{2.25}$$

3. FURTHER POSSIBLE INVESTIGATIONS

It should now be obvious that many of the results known about the Fibonacci sequence can be related to the sequence given in (2.1).   To continue this development would be a trivial pursuit.   However, there are several challenging questions which the authors would like to propose to the reader:

(1) Can one develop divisibility properties similar to those known about the Fibonacci sequence?

(2) What is the value of the largest perfect square in any one of the generalized sequences?

(3) Do triangular numbers appear more frequently in the generalized sequences than in the Fibonacci sequence?

(4) Are the sequences complete from the integer representation point of view?

(5) How are the roots of the equation $x^n = x^{n-1} + x^{n-2} + 1$ related to numbers of the sequence in (1.1)?

REFERENCES

[1] Horibe, Y. "An Entropy View of Fibonacci Trees." *The Fibonacci Quarterly,* *Vol. 20, No.* 2 (1982): pp 168-178.

[2] Horibe, Y. "Notes on Fibonacci Trees and Their Optimality." *The Fibonacci Quarterly, Vol. 21, No.* 2 (1983): pp 118-128.

Dmitri Thoro

FIRST FAILURES

Vern Hoggatt Jr. had an insatiable appetite for problem solving at all levels of mathematics. Thus, it is not surprising that he so often exhibited his contagious enthusiasm for the types of "classroom anecdotes" contained in this paper. I fondly recall his support.

In each of the anecdotes that follow you are asked to pretend that you are a naive student who has prematurely formulated a conjecture. Then look for the first value of n which <u>disproves</u> your claim. I would be delighted to receive your comments. Note that some of the descriptions are purposefully vague! But, for assistance, you will find a set of remarks following the list of conjectures.

Conjecture 1: Let $H_n = 4 + \sum_{k=1}^{n} k!$ where $n \geq 1$. The first five terms of the sequence are 5, 7, 13, 37 and 157. What is your guess?

Conjecture 2: Find a modulus m such that for $n \geq 2$, $2^n \pmod{m}$ produces the sequence of terms 0, 2, 0, 2, 4, 2, 0, 8, . . .

Conjecture 3: For $n \geq 5$, examine the base five representation of 2^n.

Conjecture 4: For $n \geq 2$, the minimum number of multiplications needed to compute x^n is given by the procedure described in Prob. B-484, <u>Fibonacci Quarterly</u> 21 (1983): p 308. How often is the answer correct?

Conjecture 5: For $n \geq 1$, find the largest power of 2 which is a divisor of $(2n)!$

Conjecture 6: For $n \geq 3$, let {a, b, c} be a subset of {1, 2, 3, . . . , n}. How many of these subsets have the property that $a + b + c$ is a multiple of 3? What is true about the terminal digits of these numbers?

207

. N. Philippou et al. (eds.), Applications of Fibonacci Numbers, 207–210.
1988 by Kluwer Academic Publishers.

Conjecture 7: In how many ways can 3 numbers be chosen from {1, 2, . . . n}, where n ≥ 4, if we require that they represent the lengths of the sides of a triangle?

Conjecture 8: What is especially interesting about the coefficients of the factors of $x^n - 1$?

Conjecture 9: In considering bichromatic graphs with n vertices, one encounters the formula

$$a_n = m(m-1)(m-2)/3 \quad \text{when } n = 2m$$
$$a_n = 2m(m-1)(4m+1)/3 \quad \text{when } n = 4m+1$$
$$a_n = 2m(m+1)(4m-1)/3 \quad \text{when } n = 4m+3.$$

What is interesting about a_n and a_{n+1}?

Conjecture 10: For n ≥ 1, record $\sqrt{24n+1}$ if it is an integer. The first 7 values are 5, 7, 11, 13, 17, 19, 23. What is your guess?

Conjecture 11: For n ≥ 1, record those integers for which $\sqrt{24n+1}$ is not an integer and which are not a part of your "natural conjecture" for Conjecture 10.

Conjecture 12: For some values of n ≥ 4, 2^n-7 suggests the numbers 3, 5, 11, 13, 181, . . .   What values of n work?

Conjecture 13: For n ≥ 1, consider those partitions of n for which no part exceeds 3; That is, 4 has 4 such partitions which are 1+1+1+1 = 1+1+2 = 1+3 = 2+2.

Conjecture 14: For n ≥ 1, determine the number of ways one can tile a 2×n checkerboard using 1×2 tiles? For example, a 2×3 board has 3 possible tilings.

Conjecture 15: The ubiquitous Catalan numbers may be defined by $T_n = \binom{2n}{n}/(n+1)$. What might be true about odd T_n, modulo 10?

Conjecture 16: For $n \geq 0$, what can be said about the left-most digit of 2^n? That is, for $n \leq 40$, no power of 2 begins with $\underline{?}$.

Conjecture 17: Let $f(n) = 2n^2 + 29$; then $f(n)$ is a prime for $-28 \leq n \leq 28$. If a prime p is a divisor of $f(n)$ for some n, then what can be said about $(p-29)/2$?

Conjecture 18: In Conjecture 19, we observe that if $p \equiv 1$ (mod 8), or $p \equiv 5$ (mod 8) and $(p-29)/2 \neq$ square, or $p \equiv 7$ (mod 8) and $(p-29)/2 \neq$ square then p is never a factor of $f(n)$. Is this always true?

Conjecture 19: Given n consecutive positive integers, there is one of these which is relatively prime to all the rest. The first failure is not a Fibonacci number, but one that Gauss loved! Find it.

REMARKS

Remark 1: This problem is reminiscent of the more famous problem which deals with adding 1 to the product of the first n primes.

Remark 2: Let the modulus be m. One conjecture is that the result is 0 or a power of 2. Another is that you never get 3. (For the latter, the first failure occurs when $m > 10^9$.)

Remark 3: Conjecture: at least one digit is 0. The second failure (if there is one) occurs when $n > 2^{500}$.

Remark 4: For $2 \leq n \leq 100$, the binary method fails to provide the minimum number desired in exactly 1/3 of the cases.

Remark 5: The first four values are 1, 3, 4, 7. Finding a formula for the largest power of a prime p which divides n! establishes a famous result due to Legendre. Note that there is an interesting connection between Conjectures 4 and 5. (Hint: use $[\log_2 n]$.)

Remark 6: Once more we get the ubiquitous powers of two. Furthermore, 9 is never a terminal digit.

Remark 7: Find several formulas for the general solution. An especially compact formulation is desirable.

Remark 8: Conjecture: for any factor, each coefficient is 0, 1 or -1. The first failure occurs for n between 100 and 110.

Remark 9: Consider the numbers involved modulo 10. A surprise awaits you!

Remark 10 and 11: One first failure occurs at n = 26; another at n = 610.

Remark 12: A first failure occurs at n = 39.

Remark 13 and 14: Both have similar false starts, but Conjecture 14 yields old friends.

Remark 15: One first failure occurs at n = 255.

Remark 16: Consider something like every ? consecutive powers of 2 contain two beginning with j, $1 \leq j \leq 9$; ? is minimal.

Remark 17: Similar results hold for Euler's infamous polynomial $n^2 + n + 41$.

Remark 18: Several first failures occur for n between 200 and 300.

Remark 19: Gauss wanted this number engraved on his tombstone.

SUBJECT INDEX

Asveld's Polynomials	163
Base Five	207
Bernoulli Trials	149
Bessel Function	7
Bichromatic Graphs	208
Binary Method	209
Binet's Formulas	194
Binomial Equation	1
Catalan Numbers	209
Characteristic Polynomial	39
Computational Examples	155
Congruence Relation	39
Consecutive-k-out-of-n-F: System	150
Consecutive Successes	93, 150
Coset Enumeration	55
Covering, Disjoint	144
Covering, Regular	143
Cyclically Presented Group	47
Defective Recurrence	113
Deficiency (of group)	46
Deficiency (of presentation)	46
Determinant	199
Difference Equation	39
Diophantine Equation	1
Diophantus	177, 183
Euler's Polynomial	210
Exponent of the Multiplier	116
Factorials	207
F-Representation; F-Addends in the Sum Representation	97
Fermat	177
Fermat's Last Theorem	1

Fermat's Theorem   42
Fibonacci Coefficients   66
Fibonacci Group   47
Fibonacci Identities   73
Fibonacci Number   1, 39, 49, 77, 163, 180, 183
Fibonacci Polynomials of Order k   89
Fibonacci Row   1
Fibonacci-Type Polynomials   149
Fibonacci-Type Polynomials of Order k   89
First   153
Functional Equation   19
Functions of Matrices   71
Gauss   209
Generalized Fibonacci Group   48
Generalized Fibonacci Numbers   163, 193
Generating Function   90, 194
Homogeneous Part   164
Integer-Solutions   149
Kronecker Product of Matrices   69
L-Representation; L-Addends in the Sum Representation   97
Legendre   210
Limit Theorems   201
Linear Recurrence of Order k   89
Longest   149
Longest Failure Run   153
Lucas Number   29, 39, 47, 77, 184
Lucas Sequence of the First Kind   115
Lucas Sequence of the Second Kind   119
Modified Todd-Coxeter   55
Multinomial Coefficients   61
Multinomial Expansion   90
Multinomial Theorem   40
Multiplier   116
Particular Solution   164

Partitions   208
Pascal's Triangle   1, 61
Pell Numbers   167
Period of Recurrence   115
Polynacci Polynomials of Order k   89
Presentation (for a group)   45
Prime Power Divisors   29
Primitive Divisors   29
Radic Valuation   62
Rank of Apparition   115
Rayleigh Function   8
Rayleigh Polynomial   14
Recurrence   149
Recurrence Relation   39
Recursion Relation   29
Recursive   149
Recursive Behavior, Pattern, Nature   98
Reliability   149
Restricted Period   115
Solutions in Natural Numbers   1
Special Recurrence Sequence   19
Success Run   150
Summations   197
Symmetric Recursive Sequence   17
Uniform Distribution in $(Z/mZ)^*$ 18
Weakly Uniform Distribution mod m   18
Zeckendorf's Theorem   97

Fibonacci Numbers and Their Applications

Edited by

ANDREAS N. PHILIPPOU
University of Patras, Patras, Greece

GERALD E. BERGUM
University of South Dakota, Brookings, South Dakota, U.S.A.

and

ALWYN F. HORADAM
University of New England, Armidale, N.S.W., Australia

This book describes recent advances within the field of elementary number theory (Fibonacci Numbers) and probability theory, and furnishes some novel applications in electrical engineering (ladder networks, transmission lines) and chemistry (aromatic hydrocarbons).

Fibonacci Numbers have played a dominant role in the expansion of many branches of science and engineering, with emphasis on optimization theory and computer science. The papers presented in this volume demonstrate that the applications of Fibonacci Numbers are far more extensive than may be commonly realised.

The book contains reviewed papers arising from the 'First International Conference on Fibonacci Numbers and Their Applications'.

ISBN 90–277–2234–X   MA 28

TABLE OF CONTENTS

EDITOR'S PREFACE
REPORT
CONTRIBUTORS
FOREWORD
THE ORGANIZING COMMITTEES
LIST OF CONTRIBUTORS TO THE CONFERENCE
INTRODUCTION

FIBONACCENE
  Peter G. Anderson

ON A CLASS OF NUMBERS RELATED TO BOTH THE FIBONACCI
AND PELL NUMBERS
  Nguyen-Huu Bong

A PROPERTY OF UNIT DIGITS OF FIBONACCI NUMBERS
  Herta T. Freitag

SOME PROPERTIES OF THE DISTRIBUTIONS OF ORDER k
  Katuomi Hirano

CONVOLUTIONS FOR PELL POLYNOMIALS
  A. F. Horadam & Br. J. M. Mahon

CYCLOTOMY-GENERATED POLYNOMIALS OF FIBONACCI TYPE
  A. F. Horadam & A. G. Shannon

ON GENERALIZED FIBONACCI PROCESS
  Daniela Jarušková

FIBONACCI NUMBERS OF GRAPHS III: PLANTED PLANE TREES
  Peter Kirschenhofer, Helmut Prodinger,
  & Robert F. Tichy

A DISTRIBUTION PROPERTY OF SECOND-ORDER LINEAR RECURRENCES
  Péter Kiss

ON LUCAS PSEUDOPRIMES WHICH ARE PRODUCTS OF s PRIMES
  Péter Kiss, Bui Minh Phong, & Erik Lieuwens

FIBONACCI AND LUCAS NUMBERS AND THE MORGAN-VOYCE POLY-
NOMIALS IN LADDER NETWORKS AND IN ELECTRIC LINE THEORY
  Joseph Lahr

INFINITE SERIES SUMMATION INVOLVING RECIPROCALS OF
PELL POLYNOMIALS
  Br. J. M. Mahon & A. F. Horadam

FIBONACCI AND LUCAS NUMBERS AND AITKEN ACCELERATION
  J. H. McCabe & G. M. Phillips

ON SEQUENCES HAVING THIRD-ORDER RECURRENCE RELATIONS
  S. Pethe

ON THE SOLUTION OF THE EQUATION $G_n = P(x)$
  Attila Pethö

DISTRIBUTIONS AND FIBONACCI POLYNOMIALS OF ORDER k,
LONGEST RUNS, AND RELIABILITY OF CONSECUTIVE-k-OUT-
OF-$n : F$ SYSTEMS
  Andreas N. Philippou

FIBONACCI-TYPE POLYNOMIALS AND PASCAL TRIANGLES
OF ORDER k
  G. N. Philippou & C. Georghiou

A NOTE ON FIBONACCI AND RELATED NUMBERS IN THE THEORY
OF 2×2 MATRICES
  Gerhard Rosenberger

PROBLEMS ON FIBONACCI NUMBERS AND THEIR GENERALIZATIONS
  A. Rotkiewicz

LINEAR RECURRENCES HAVING ALMOST ALL PRIMES AS
MAXIMAL DIVISORS
  Lawrence Somer

ON THE ASYMPTOTIC DISTRIBUTION OF LINEAR RECURRENCE
SEQUENCES
  Robert F. Tichy

GOLDEN HOPS AROUND A CIRCLE
  Tony van Ravenstein, Keith Tognetti,
  & Graham Winley